INSTANT ANATOMY

Robert H. Whitaker
MA, MD, MChir, FRCS, FMAA
Selwyn College
University of Cambridge

Neil R. Borley
MB, BS, FRCS, FRCS (ed)
Cheltenham General Hospital, Cheltenham

FOURTH EDITION

WILEY-BLACKWELL
A John Wiley & Sons, Ltd., Publication

This edition first published 2010, © 2010 by Robert H. Whitaker and Neil R. Borley
Previous editions: 1994, 2000, 2005

Blackwell Publishing was acquired by John Wiley & Sons in February 2007. Blackwell's publishing program has been merged with Wiley's global Scientific, Technical and Medical business to form Wiley-Blackwell.

Registered office: John Wiley & Sons Ltd, The Atrium, Southern Gate, Chichester, West Sussex, PO19 8SQ, UK

Editorial offices: 9600 Garsington Road, Oxford, OX4 2DQ, UK
The Atrium, Southern Gate, Chichester, West Sussex, PO19 8SQ, UK
111 River Street, Hoboken, NJ 07030-5774, USA

For details of our global editorial offices, for customer services and for information about how to apply for permission to reuse the copyright material in this book please see our website at www.wiley.com/wiley-blackwell

Library of Congress Cataloging-in-Publication Data

Whitaker, R. H. (Robert H.)
Instant anatomy / Robert H. Whitaker, Neil R. Borley. — 4th ed.
p. ; cm.
Includes bibliographical references.
ISBN 978-1-4051-9961-2
1. Human anatomy—Outlines, syllabi, etc. I. Borley, Neil R. II. Title.
[DNLM: 1. Anatomy—Outlines. QS 18.2 W578i 2010]
QM31.W55 2010
612—dc22

2009046386

ISBN: 978-1-4051-9961-2

A catalogue record for this book is available from the British Library.

Set in 8.5/10.5pt Sabon by Graphicraft Limited, Hong Kong
Printed in Singapore

1 2010

THE AUTHORS

Robert H. Whitaker MD, MChir, FRCS, FMAA graduated from the University of Cambridge and trained at University College Hospital, London. He spent a year at Johns Hopkins Hospital, Baltimore, in the Urological Research Laboratories before returning to continue his training first at the St Peters Hospital group in London then as a Senior Lecturer in Urology at the London Hospital Medical School. He was appointed as a Consultant Urologist at Addenbrooke's Hospital in Cambridge in 1973 and spent 20 years practising mostly paediatric urology before retiring from clinical practice to join the Department of Anatomy in Cambridge to help with the teaching of students and trainee surgeons.

Neil R. Borley MB, BS, FRCS, MS trained at Guy's Hospital, London. He undertook a surgical rotation at Addenbrooke's Hospital, Cambridge, before becoming a Demonstrator in the Department of Anatomy in Cambridge under Professor Harold Ellis. He passed the Primary FRCS examination in 1993 for which he received the Hallet Prize and then continued his surgical training at Papworth Hospital, Kent & Canterbury Hospital and moved to Oxford as Surgical Registrar and then Clinical Lecturer in Surgery and Clinical Tutor in the Nuffield Department of Surgery. He is now a Consultant Colorectal Surgeon in Cheltenham.

CONTENTS

PREFACE TO FOURTH EDITION

On reading again the prefaces to the first three editions it seems to us that we have achieved our objectives since the first one was written in 1994. We have preserved the book for quick reference and user friendliness. Any additions or changes over the years appear to have become popular.

We continue to believe that anatomy is the language of medicine and that at qualification as a doctor, physiotherapist, nurse or radiographer, to name just a few professions, there is a basic amount of anatomical knowledge that is essential. To us, it has become ominously clear that there has been a continued and substantial move away from this emphasis on 'need to know' anatomy in many medical schools, and even worse, in higher surgical training / examinations; in our experience the level of anatomical knowledge across specialities may have fallen to depressingly low levels. Can there be a point of no return after which these worrying trends will become irreversible? We hope not but remain much concerned.

In this fourth edition we have made several substantial changes and hope that you approve of them. You, the readers, have sent us a number of corrections or improvements for which we are eternally grateful. The section on 'Spaces other then skull and spine' has been removed and the previous contents have been illustrated and added to the 'Surface anatomy and key areas' section at the end of the book. Over the years there have been minimal demands for an index which indicates that with the colour coding and the easily found relevant sections readers are content. Despite its slight expansion it is still an instant reference book for the slightly larger white coat pocket and briefcase.

Please keep your comments coming in; as you can see we take them to heart. Also we urge you to take a look at our Instant Anatomy website (***instantanatomy.net***) where you will find a whole new set of illustrations, podcasts, questions and answers, multiple choice questions and mini-Powerpoint presentations on all aspects of anatomy.

Finally, after 16 years this is a great opportunity to thank again Jane Fallows, our faithful illustrator, and also show our appreciation of our long-lasting and excellent relationships we value with Wiley Blackwell.

ROBERT WHITAKER
NEIL BORLEY
Cambridge and Cheltenham, 2010

PREFACE TO FIRST EDITION

How many times have you looked up the course of an artery or nerve in one of the excellent anatomy textbooks that are available today only to find that the details are spread over several sections of the book and that an instant summary is not available? At times like this you wish there was a quick reference book with all the answers neatly catalogued in dictionary format.

We have attempted to provide such a concise text for rapid reference. Of course, we emphasise that this is not a text for learning anatomy from scratch but one that should be used in conjunction with one of the fuller texts that has stood the test of time. The book is designed for those who already have some working knowledge of anatomy and need to find accurate facts quickly.

Both authors have been sufficiently recent students of anatomy for higher degrees and for teaching undergraduate medical students that each can remember the problems that both students and they themselves encountered. The book has been compiled with this in mind.

It is designed primarily for undergraduate medical students and prospective surgeons who are studying for a higher degree in surgery. For each of these groups we believe it will be ideal. However, it should also be extremely useful for all clinicians who need to remind themselves of anatomical facts at all stages in their careers and for other professional groups such as nurses, physiotherapists and radiographers.

Inevitably in a book of this size there has been some selection of material for inclusion and no attempt has been made to provide details of minutiae that appear in the fuller text.

The authors' original artwork was redrawn with a graphics program by Jane Fallows, medical illustrator, to whom the authors owe an immense debt of gratitude for her skill and patience.

ROBERT WHITAKER
NEIL BORLEY
Cambridge, 1994

NOTES ON THE TEXT

The illustrations show the **right** side of the body as viewed from in front, unless otherwise indicated. The two exceptions are the cervical and brachial plexuses where it makes little difference as to which side they are viewed and they are more conveniently drawn and remembered as they are shown here. Where there might be confusion, a small compass has been added to indicate the left and right *sides of the body*.

Eponymous names appear sparingly and only when they are in common usage. The following abbreviations have been used as appropriate throughout the text.

List of abbreviations

ant	anterior(ly)
art(s)	artery(ies)
br(s)	branch(s)
CMC	carpometacarpal
div(s)	division(s)
ext	external
inf	inferior(ly)
int	internal
IP	interphalangeal
jnt(s)	joint(s)
lat	lateral(ly)
lig(s)	ligament(s)
med	medial(ly)
MC(s)	metacarpal(s)
MCP	metacarpophalangeal
MTP	metatarsophalangeal
MT(s)	metatarsal(s)
N(s) or n(s)	nerve(s)
post	posterior(ly)
prox	proximal
sup	superior(ly)
TMT	tarsometatarsal
V(s)	vein(s)

Note: Abbreviations are not used for muscle names or in titles. The following words are always written in full: greater, lesser, middle, superficial and combinations such as posterolateral.

1: ARTERIES

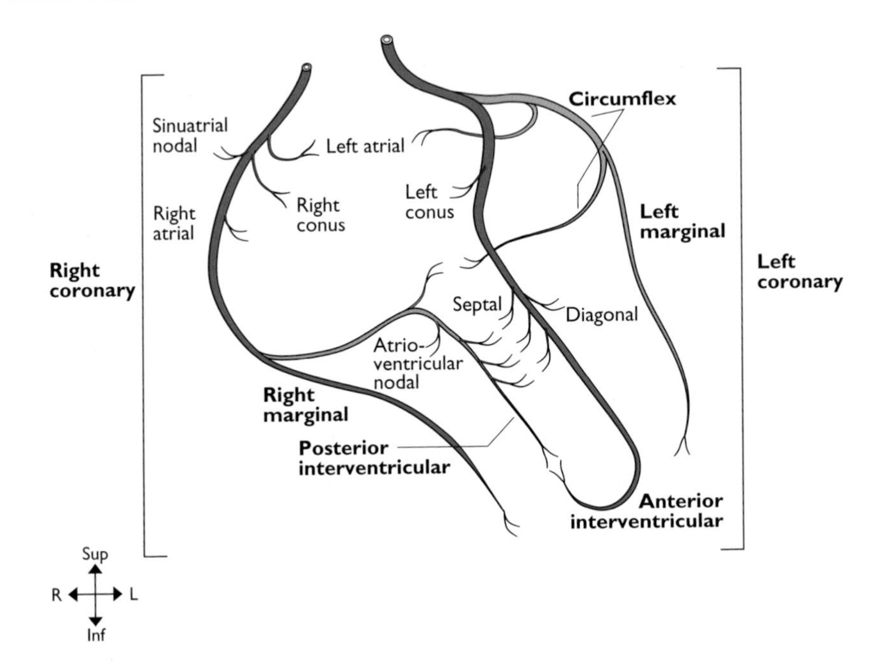

Coronary arteries

CORONARY ARTERIES

From: Ascending aorta
To: Myocardium

Right coronary artery. Originates from the anterior aortic sinus. It passes anteriorly between the pulmonary trunk and the right auricle to reach the atrioventricular sulcus in which it runs down the anterior surface of the right cardiac border and then onto the inferior surface of the heart. It terminates at the junction of the atrioventricular sulcus and the posterior interventricular groove by anastomosing with the circumflex branch of the left coronary artery and giving off the posterior interventricular (posterior descending) artery. It supplies the right atrium and part of the left atrium, the sinuatrial node in 60% of cases, the right ventricle, the posterior part of the interventricular septum and the atrioventricular node in 80% of cases.

Left coronary artery. Arises from the left posterior aortic sinus. It passes laterally, posterior to the pulmonary trunk and anterior to the left auricle to reach the atrioventricular groove where it divides into an anterior interventricular (formally left anterior descending) artery and circumflex branches. The circumflex artery runs in the atrioventricular sulcus around the left border of the heart to anastomose with the right coronary artery. The anterior interventricular artery descends on the anterior surface of the heart in the anterior interventricular groove and around the apex of the heart into the posterior interventricular groove where it anastomoses with the posterior interventricular branch of the right coronary artery. The left coronary artery supplies the left atrium, left ventricle, anterior interventricular septum, sinuatrial node in 40% of cases and the atrioventricular node in 20%.

Dominance. In approximately 10% of hearts the posterior interventricular artery arises from the circumflex artery (left coronary) and then most of the left ventricle and interventricular septum are supplied by the left coronary artery. The heart is said to have *left cardiac dominance*.

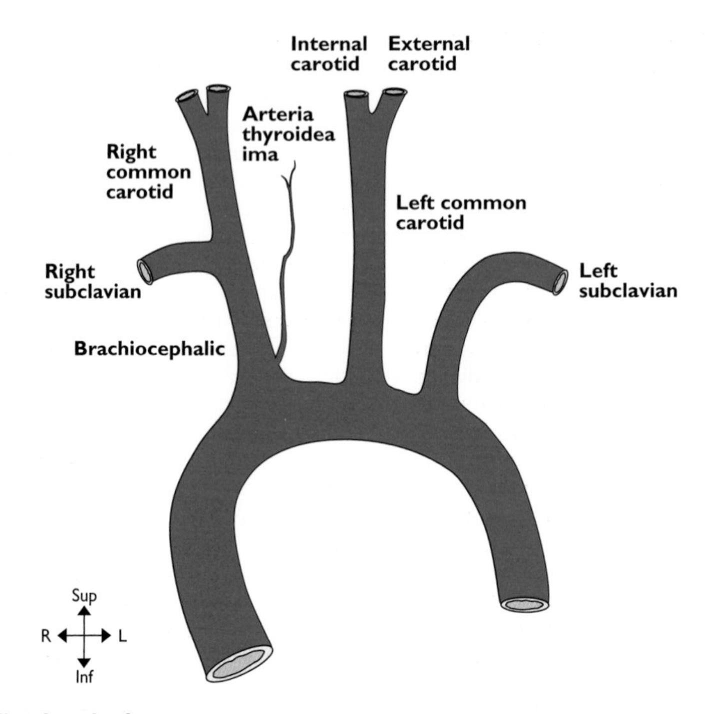

Ascending & arch of aorta

ASCENDING & ARCH OF AORTA

From: **Left ventricle**
To: **Descending aorta**

Ascending aorta. Arises at the vestibule of the left ventricle at the level of the third left costal cartilage and passes upwards and slightly to the right to a point behind the sternum at the level of the manubriosternal joint (second costal cartilage) where it becomes the arch of the aorta. It is enclosed in fibrous and serous pericardium. Anterior to it are the right auricle, the infundibulum of the right ventricle and pulmonary trunk. Posterior, lie the left atrium, the right pulmonary artery and right main bronchus. To the left lie the pulmonary trunk and the left auricle. To the right are the superior vena cava and the right atrium.

Arch of aorta. The arch begins posterior to the manubriosternal joint at the level of the second costal cartilage and passes posterior and to the left, over the left main bronchus to end at the left side of the body of T4 vertebra. Its highest level is the mid point of the manubrium sterni and at this level its three main branches emerge. Anterior and to the left of the arch are (from anterior to posterior) the left phrenic nerve, vagal and sympathetic contributions to the cardiac plexus, and the left vagus. Also, the left superior intercostal vein runs forwards on the arch anterior to the vagus and posterior to the phrenic nerve. Lateral to all these structures are the pleura and left lung. Posterior and to the right of the arch are the trachea, deep cardiac plexus, left recurrent laryngeal nerve, oesophagus, thoracic duct and the body of T4. Inferior to the arch are the pulmonary bifurcation, the left main bronchus, the ligamentum arteriosum and the left recurrent laryngeal nerve. From its superior surface emerge the brachiocephalic artery, the left common carotid and left subclavian arteries. Within the adventitia of the ascending and arch of the aorta lie baro- and chemoreceptors.

Brachiocephalic artery. Arises from the convexity of the aortic arch behind the manubrium sterni and passes upwards and posteriorly to the right. It divides into the right subclavian and right common carotid arteries posterior to the right sternoclavicular joint. Anterior to it are the left brachiocephalic vein with the right inferior thyroid vein entering it, and the thymic remnants. The artery initially lies anterior to the trachea then passes to lie on its right lateral side. On the right of the artery are the right brachiocephalic vein, upper part of the superior vena cava, the pleura and the cardiac branches of the vagus. The main vagal trunk is more posterolateral. At the origin of the brachiocephalic artery the left common carotid artery lies posteriorly on its left.

continued

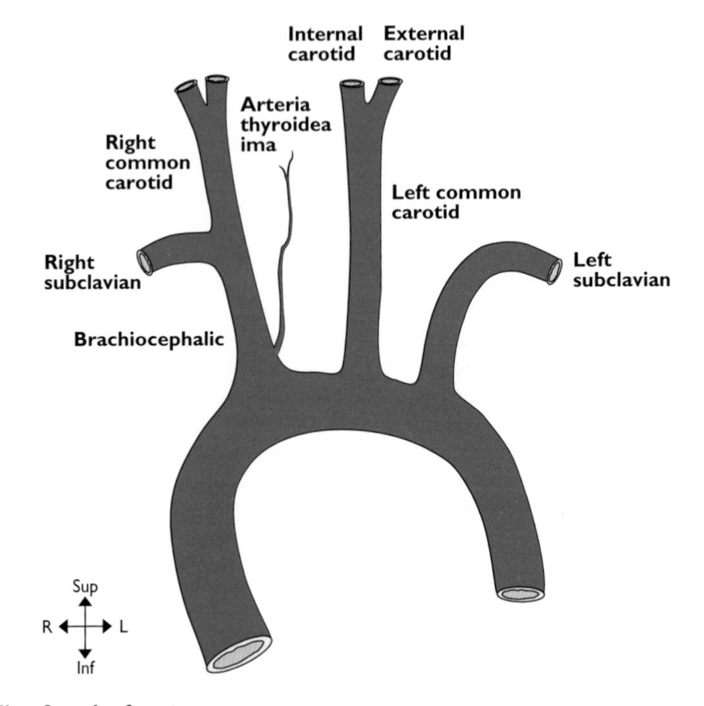

Ascending & arch of aorta

Common carotid arteries. The right common carotid artery arises from the brachiocephalic artery as it divides posterior to the right sternoclavicular joint, whilst the left common carotid arises from the convexity of the aortic arch. Both end as the arteries bifurcate at the level of the upper border of the thyroid cartilage (C4).

Left common carotid artery (thorax). Lying anterior to the thoracic part of this artery are the left brachiocephalic vein and the thymic remnant. Posterior to it in its lower part are the left subclavian artery and the trachea whilst further superiorly there is the left recurrent laryngeal nerve, the thoracic duct and the left side of the oesophagus. On its right at its origin is the brachiocephalic artery but as it ascends the inferior thyroid veins and the trachea come to lie on its right side. To its left lie the vagus, the left phrenic nerve and the left pleura and lung.

Both common carotid arteries (cervical). Ascend in the neck slightly laterally from a point posterior to the sternoclavicular joint to end at the level of the upper border of the thyroid cartilage (C4) at which point there is a dilatation—the carotid sinus (a baroreceptor). On the posterior aspect of the bifurcation there is the carotid body (a chemoreceptor). Lying between left and right arteries, and medial to each, progressively from below are the trachea, recurrent laryngeal nerves, thyroid gland, larynx and pharynx. Each artery lies in its carotid sheath with the internal jugular vein lateral to it and the vagus nerve between and posterior to them both.

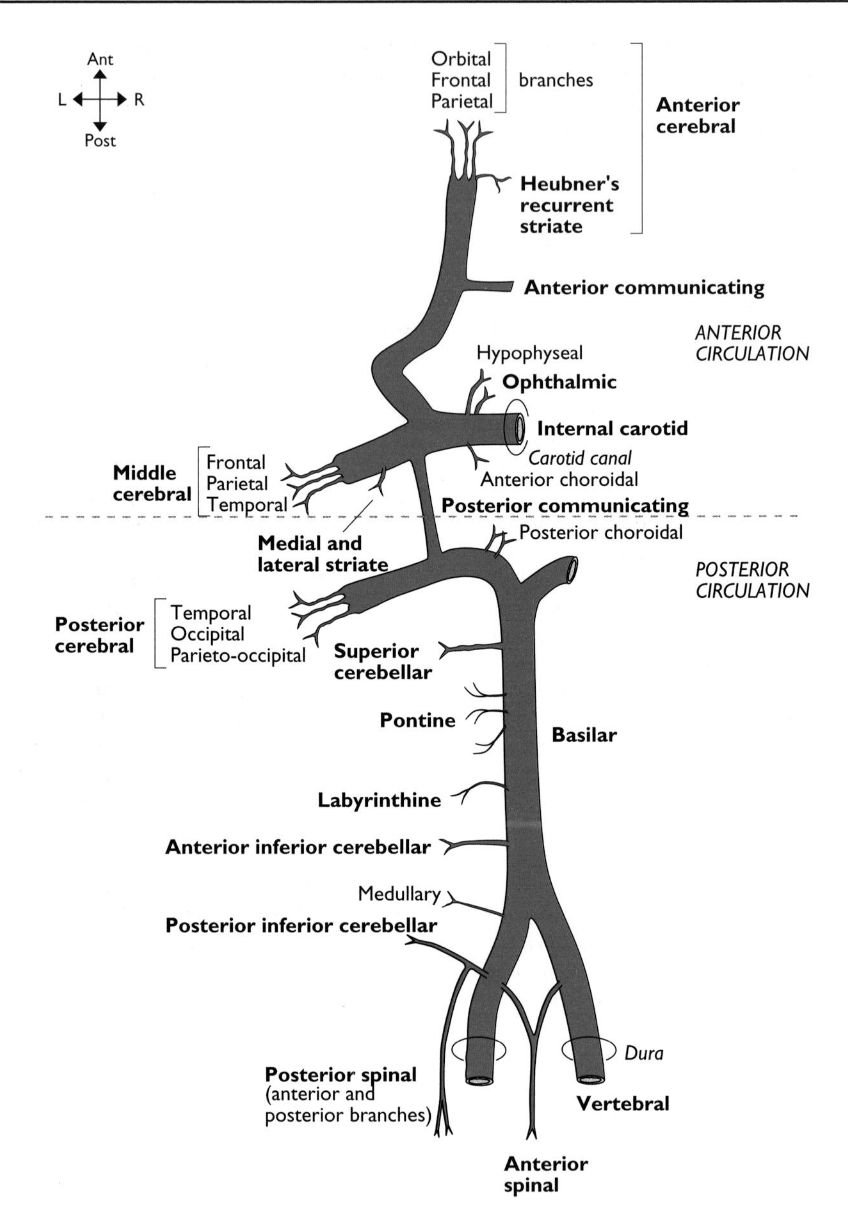

Internal carotid, vertebrobasilar system & circle of Willis

Note: (1) Labyrinthine usually arises from anterior inferior cerebellar; (2) posterior spinal may come from vertebral

INTERNAL CAROTID ARTERY, VERTEBROBASILAR SYSTEM & CIRCLE OF WILLIS

From: **Bifurcation of the common carotid arts (C4) & first parts of subclavian arts**
To: **Terminal brs**

The internal carotid artery angles from the bifurcation slightly posteriorly to reach the carotid canal through which it enters the skull to end as middle and anterior cerebral arteries. At its origin it possesses a dilatation in which lie the carotid sinus and body.
In the neck it is crossed laterally by, from below up, the pharyngeal branch of the vagus (X), glossopharyngeal nerve (IX), stylopharyngeus and styloglossus. It lies on the pharyngeal wall and the pharyngobasilar fascia. Within the carotid canal it turns 90 degrees anteromedially to run through the petrous temporal bone where it lies medial to the middle ear. It then turns 90 degrees superiorly to pass across the upper limit of the foramen lacerum. It then turns 90 degrees anteriorly to pass forwards, lateral to the body of the sphenoid which it grooves. Here it lies in the medial wall of the cavernous sinus with the abducent nerve (VI) on its lateral side. At the anterior end of the cavernous sinus it turns 90 degrees superiorly then 90 degrees posteriorly to pass medial to the anterior clinoid process and lateral to the pituitary stalk and optic chiasma. It ends as terminal branches on the medial surface of the temporal lobe.

Anterior cerebral artery is formed by the bifurcation of the internal carotid artery. It passes anteriorly over the optic nerve to arch over the genu of the corpus callosum on the medial aspect of the cerebral hemispheres where it ends as terminal branches.

Middle cerebral artery is formed by the bifurcation of the internal carotid artery. It runs laterally into the sylvian fissure then posterosuperiorly in the sulcus where it divides into terminal branches.

Basilar artery is formed by the junction of the left and right vertebral arteries (see subclavian artery, pp. 16–19) anterior to the upper medulla. From there it ascends lying angled forwards between the pons and the clivus in a slight depression on the anterior surface of the pons. It terminates at the upper border of the pons as posterior cerebral arteries.

Posterior cerebral artery is formed by the bifurcation of the basilar artery. It passes laterally around the cerebral peduncle to run posteriorly above the tentorium cerebelli on the inferomedial surface of the occipital lobe where it divides into terminal branches.

(Other branches of the internal carotid artery, not illustrated, are caroticotympanic, pterygoid and cavernous arteries.)

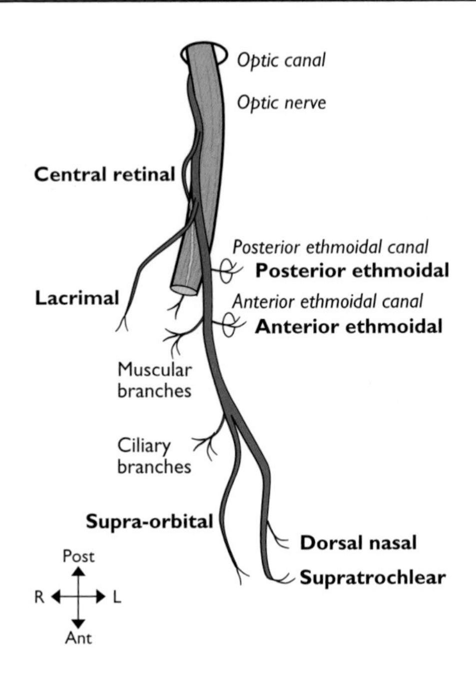

Ophthalmic artery
Note: Right side viewed from above

OPHTHALMIC ARTERY
From: **Internal carotid art**
To: **Terminal brs in orbit**

It arises from the internal carotid artery as it lies medial to the anterior clinoid process and runs anteriorly through the optic canal within the optic nerve's dural sheath, lying inferolateral to the nerve. Small branches supply the proximal nerve. In the orbit the artery leaves the dural sheath and passes forwards around the lateral side of the nerve to cross anterior to it to reach the medial orbit. It then continues medially between superior oblique and medial rectus to pass out of the cone of muscles to reach the medial wall of the orbit. The artery continues forwards to terminate at the medial orbital border deep to the superior tarsal plate as branches which leave the orbit to anastomose with branches of the facial artery.

Central retinal artery. This small, important end artery supplies the optic nerve and retina. It leaves the ophthalmic artery below the optic nerve and then, half way along the orbital part of the optic nerve, enters first the dural sheath and then the nerve itself.

(Other branches, not illustrated, (1) of ophthalmic artery are anterior meningeal and medial palpebral arteries; (2) of lacrimal artery are lateral palpebral, zygomatic and recurrent meningeal arteries; (3) of muscular is anterior ciliary artery.)

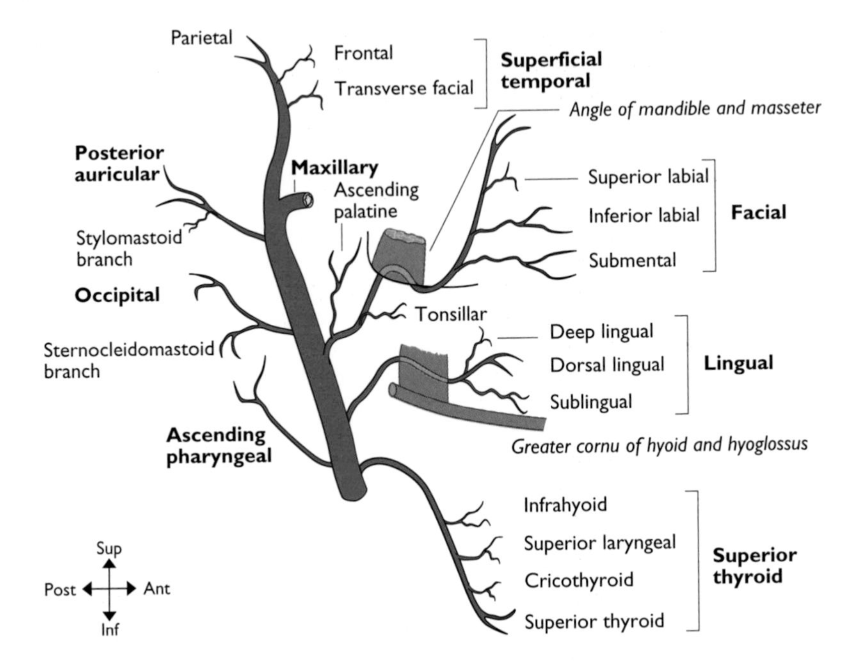

External carotid artery

EXTERNAL CAROTID ARTERY

From: **Upper border of thyroid cartilage** (C4)
To: **Terminal brs within parotid gland post to neck of mandible**

The artery arises within the carotid sheath from the bifurcation of the common carotid artery. It lies at first anteromedial to the internal carotid artery but spirals over it to come to lie lateral to it at the level of C2. Initially, it angles slightly forwards then curves backwards as it ascends to enter the parotid gland between deep and superficial lobes. During its course it is crossed by, from below upwards: the upper root of the ansa cervicalis, the hypoglossal nerve, the posterior belly of digastric, stylohyoid, the stylohyoid ligament and the facial nerve (within the parotid). Passing between it and the internal carotid artery are, from below upwards, the pharyngeal branch of the vagus (X), glossopharyngeal nerve (IX), stylopharyngeus and styloglossus. It lies on, from below upwards, pharyngeal wall, superior laryngeal branch of the vagus (X) and deep parotid lobe.

Superior thyroid artery. Arises from the anterior surface of the external carotid artery near its origin and runs inferiorly and forwards deep to omohyoid and lateral to the inferior constrictor and external laryngeal nerve to reach the upper pole of the thyroid gland.

Lingual artery. Runs superiorly looping over the greater cornu of the hyoid bone and passes medially (deep) to hyoglossus then into the substance of the tongue.

Facial artery. Arises from the anteromedial surface of the external carotid artery and runs above the hyoid bone deep to digastric and passes upwards to reach the posterior surface of the submandibular gland which it grooves deeply, lying medial to the body of the mandible. Here it lies on superior constrictor, directly lateral to the palatine tonsil. It then follows a tortuous course looping at first inferiorly then upwards around the lower border of the mandible to cross the bone anterior to the insertion of masseter (where it is easily palpable). It then runs in the superficial tissues of the face towards the angle of the mouth where it turns superiorly towards the medial canthus of the eye. (Other branches, not illustrated, are glandular (to submandibular gland) and lateral nasal arteries.)

Superficial temporal artery. Runs superiorly between the deep and superficial lobes of the parotid gland, over the posterior end of the zygomatic process (where it is easily palpable) and terminates in the subcutaneous tissues of the lateral scalp.

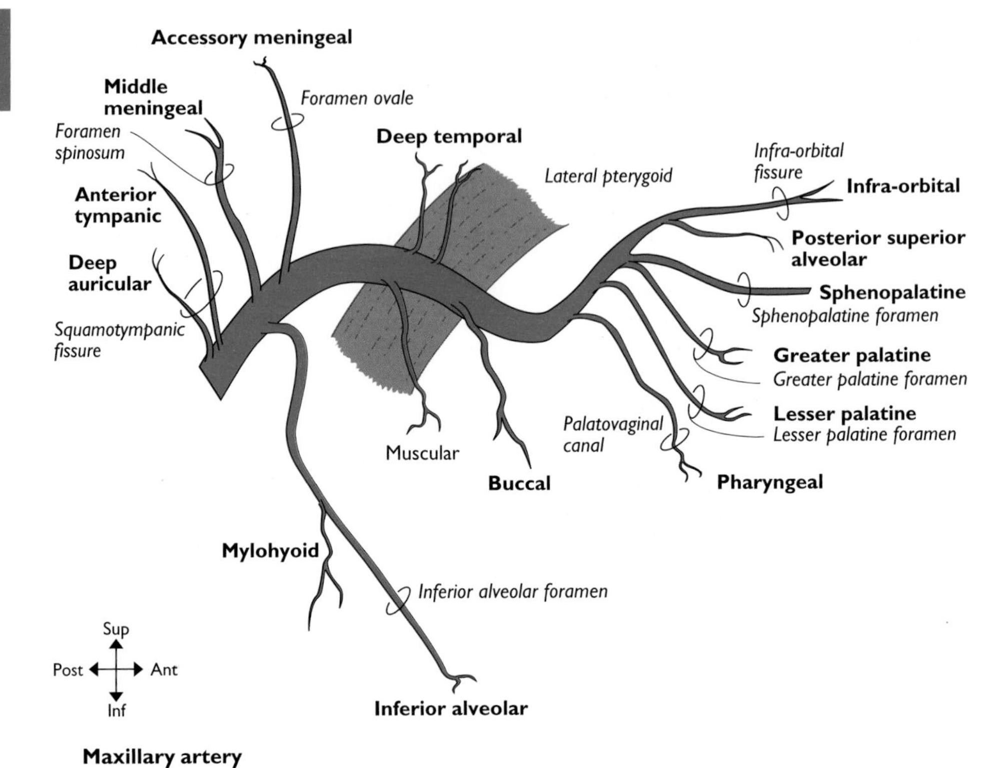

Maxillary artery

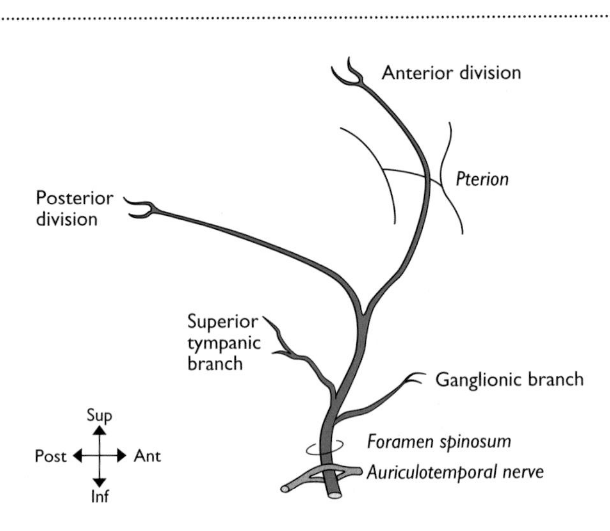

Middle meningeal artery

MAXILLARY ARTERY
From: **External carotid within parotid gland**
To: **Terminal brs in pterygopalatine fossa**

It arises from the external carotid artery within the parotid gland posterior to the neck of the mandible and ends as the sphenopalatine artery. The artery is divided into three portions by its relationship posterior, in, or anterior to the lateral pterygoid muscle. The first part passes deep to the neck of the mandible between the bone and the sphenomandibular ligament and runs anteriorly lateral to the inferior alveolar nerve to reach the border of the lateral pterygoid. The second part angles anteromedially between the two heads of lateral pterygoid between anterior and posterior divisions of the mandibular nerve. The third part leaves the lateral pterygoid to enter the pterygopalatine fossa where it terminates as branches which accompany the branches of the maxillary division of the trigeminal nerve (Vb).

Inferior alveolar artery. Passes inferolaterally posterior to the inferior alveolar nerve onto the medial surface of the ramus of the mandible which it grooves as it enters the inferior alveolar (mandibular) foramen in the mandible. It is distributed along the mandibular canal to the lower jaw and teeth. Its terminal branch appears as the mental branch through the mental foramen.

(Other branches, not illustrated, (1) of maxillary artery (third part) is artery of pterygoid canal; (2) of inferior alveolar artery are dental and mental; (3) of infraorbital artery are dental and anterior superior alveolar; (4) of posterior superior alveolar artery is dental.)

MIDDLE MENINGEAL ARTERY
From: **First part of maxillary art**
To: **Terminal brs**

It arises from the superomedial surface of the maxillary (first part) to run between the two rootlets of the auriculotemporal nerve as it passes vertically into the foramen spinosum in the greater wing of the sphenoid bone. After a very short course laterally over the greater wing of the sphenoid in the middle cranial fossa it divides into anterior and posterior divisions. The anterior division runs anterolaterally on the floor of the middle cranial fossa beneath the dura mater and grooves the greater wing of the sphenoid as it passes upwards to the junction of the lesser and greater wings. Here it may groove deeply or tunnel through the bone at the apex of the greater wing. It passes across the inner aspect of the pterion onto the parietal bone. The posterior division runs almost horizontally posterolateral over the inner aspect of the squamous temporal bone onto the lower parietal bone where it gives terminal branches.

Because of the problem of extradural haemorrhage caused by damage to this artery, the surface anatomy is important. Anterior branch: 3 cm above mid point of zygomatic arch. Posterior branch: on a line vertically from the mastoid process and horizontal from the upper margin of the orbit.

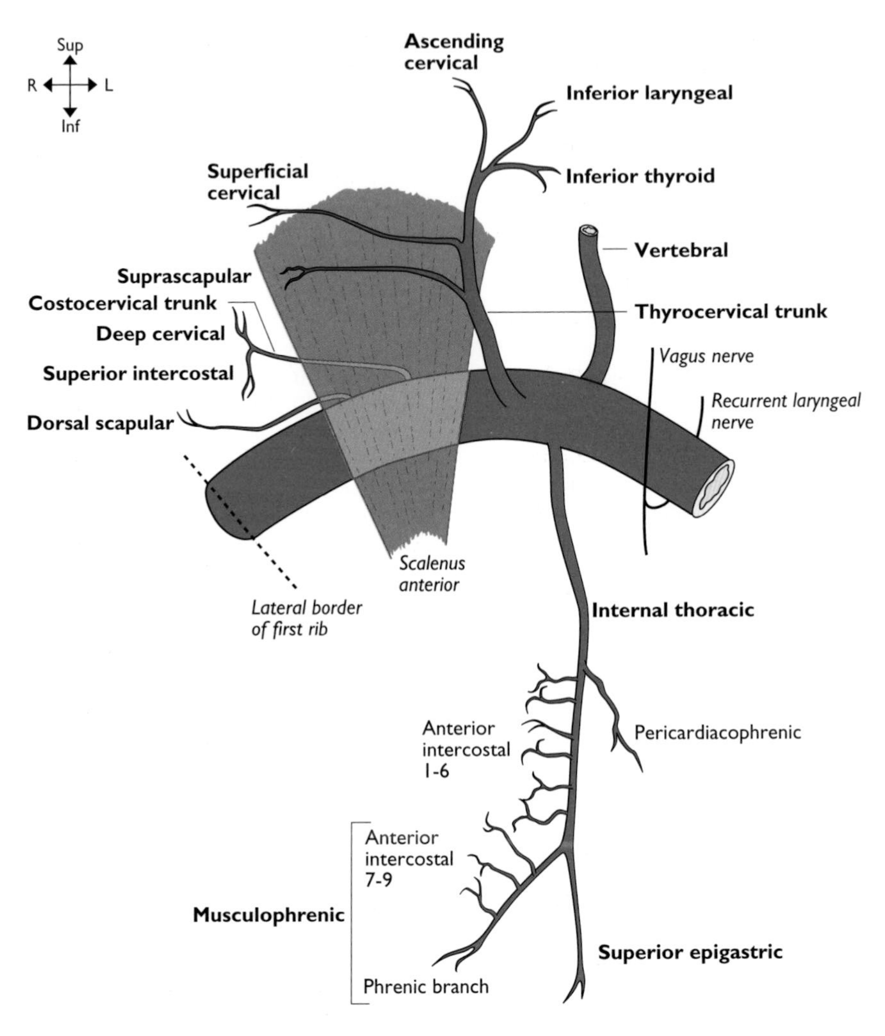

Subclavian artery

Note: (1) The superficial cervical artery is named 'transverse cervical artery' if it gives origin to the dorsal scapular artery instead of the latter arising separately from the second part of the subclavian artery; (2) phrenic branch of musculophrenic artery anastomoses with inferior phrenic artery

SUBCLAVIAN ARTERY

From: **Right—brachiocephalic trunk**
Left—aortic arch
To: **Axillary art**

The subclavian arteries arise as indicated above and end at the outer border of the first rib where they become the axillary arteries. They each have three parts: (1) medial (three branches); (2) behind (two branches); and (3) lateral (no branches) to scalenus anterior.

Right subclavian artery—first part. Arises from the brachiocephalic artery behind the right sternoclavicular joint, lying initially posterior to the right common carotid artery, then passing upwards and laterally to reach the medial side of scalenus anterior. Anterior to this first part are the vagus (X), its cardiac branches, sympathetic nerves, the internal jugular and vertebral veins. The ansa subclavia (sympathetic nerves) curls around the artery to lie both anterior and posterior to it. As the artery arches laterally the suprapleural membrane and the right recurrent laryngeal nerve lie inferior and posterior to it.

Left subclavian artery—first part. Arises from the arch of the aorta just posterior and slightly to the left of the origin of the left common carotid artery at the level of the intervertebral disc of T3/T4. It passes upwards and then, behind the left sterno-clavicular joint, it arches laterally over the suprapleural membrane to the medial edge of scalenus anterior. Anterior to it in the thorax are the left common carotid artery, the left brachiocephalic vein, the left vagus and its cardiac branches and the left phrenic nerve. Posterior to it lie the left side of the oesophagus, the thoracic duct and longus colli. Medial to it is the trachea, the left recurrent laryngeal nerve and, more superiorly, the thoracic duct. In the neck it is crossed anteriorly by the left phrenic nerve and the thoracic duct.

Subclavian artery—second part. Lies posterior to scalenus anterior and anterior to scalenus medius. Anterior to scalenus anterior are the phrenic nerve and, slightly inferior, the subclavian vein. Postero-inferior are the suprapleural membrane and the lower trunk of the brachial plexus. Superior to it are the upper and middle trunks of the brachial plexus.

Subclavian artery—third part. Begins at the lateral margin of scalenus anterior and extends to the outer (lateral) margin of the first rib where it becomes the axillary artery. Anterior to it is the external jugular vein and its tributaries. Antero-inferior is the subclavian vein. Postero-inferior is the lower trunk of the brachial plexus and the first rib. Posterosuperior are the upper and middle trunks of the brachial plexus.

Vertebral artery (see also internal carotid, vertebrobasilar system & circle of Willis, pp. 8–9). Arises from the posterosuperior aspect of the first part of the subclavian artery and ends where the arteries from the two sides join to form the basilar artery at the lower pons. It angles posteriorly between the medial border of scalenus anterior and the lateral border of longus colli in the apex of the pyramidal space before entering the foramen in the transverse process of C6 behind its anterior tubercle (carotid tubercle of Chassaignac). Lying anterior to this first part are the common carotid artery and the vertebral vein and, more medially, the inferior thyroid artery and middle cervical ganglion. On the left the thoracic duct crosses it anteriorly. Posterior to it are the anterior primary rami of C7 and C8 nerves and more medially the inferior cervical (stellate) ganglion. The second part of the artery ascends within the foramina of the transverse processes of C6 to C1, accompanied by sympathetic nerves and vertebral veins. It passes out posteriorly behind the lateral mass of the atlas before

continued

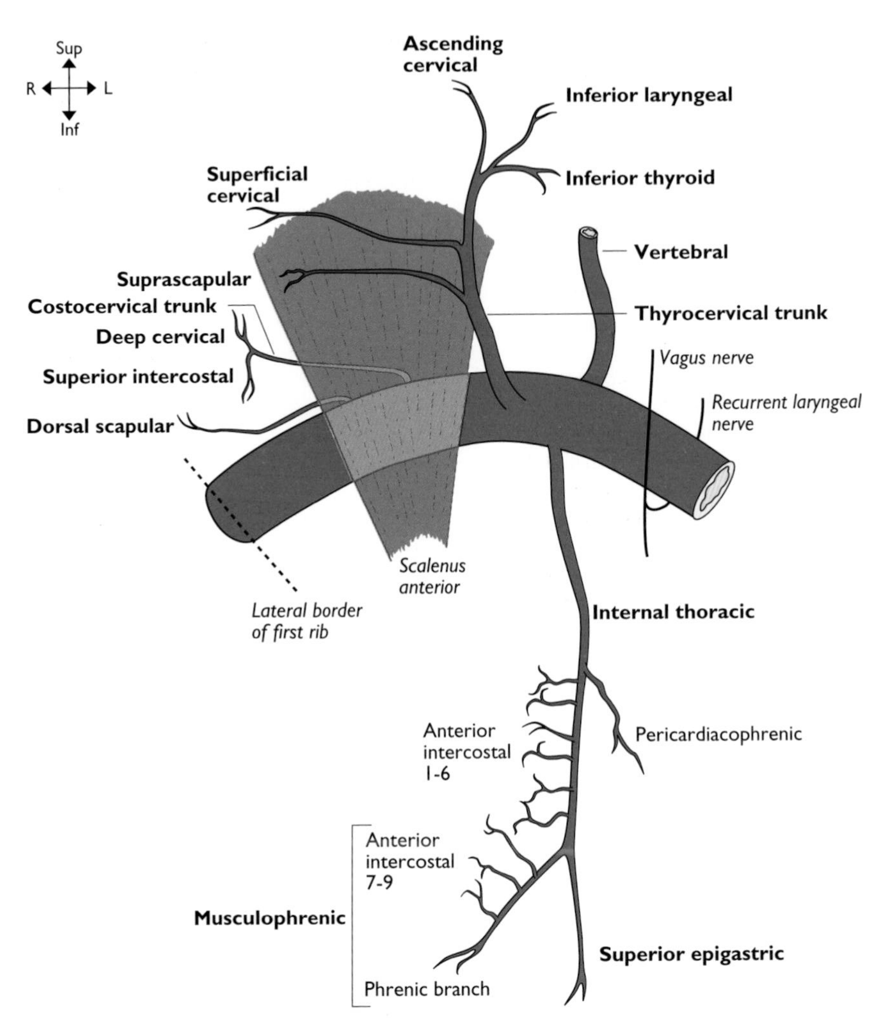

Subclavian artery

Note: (1) The superficial cervical artery is named 'transverse cervical artery' if it gives origin to the dorsal scapular artery instead of the latter arising separately from the second part of the subclavian artery; (2) phrenic branch of musculophrenic artery anastomoses with inferior phrenic artery

turning medially over its posterior arch. It then turns anteriorly to pierce the atlanto-occipital membrane lateral to the cervicomedullary junction. It pierces the dura and arachnoid to ascend supero-medially around the anterior aspect of the medulla where it joins the artery from the opposite side at the lower border of the pons to form the basilar artery. (Other branches, not illustrated, are spinal, meningeal and muscular.)

Internal thoracic artery. Arises from the anterior aspect of the first part of the subclavian artery and passes inferiorly behind the brachiocephalic vein and the phrenic nerve to reach the dome of the pleura. It then angles medially to lie posterior to the upper six costal cartilages, between the internal intercostal and transversus thoracis muscles. It terminates at the 6th intercostal space to give the superior epigastric and musculophrenic arteries. (Other branches, not illustrated, are mediastinal, thymic, sternal and perforating (mammary).)

Inferior thyroid artery. Ascends along the medial edge of scalenus anterior. Just below the anterior tubercle of C6 it turns medially to reach the lower thyroid gland, passing between vertebral artery and vein (posteriorly) and carotid sheath and sympathetic chain (anteriorly). Its terminal branches are often amongst the recurrent laryngeal nerve. (Other branches of inferior thyroid artery, not illustrated, are glandular, pharyngeal, oesophageal and tracheal.)

Superior intercostal artery. Passes inferiorly, anterior to the necks of the first two ribs to provide the posterior intercostal arteries for the first two intercostal spaces.

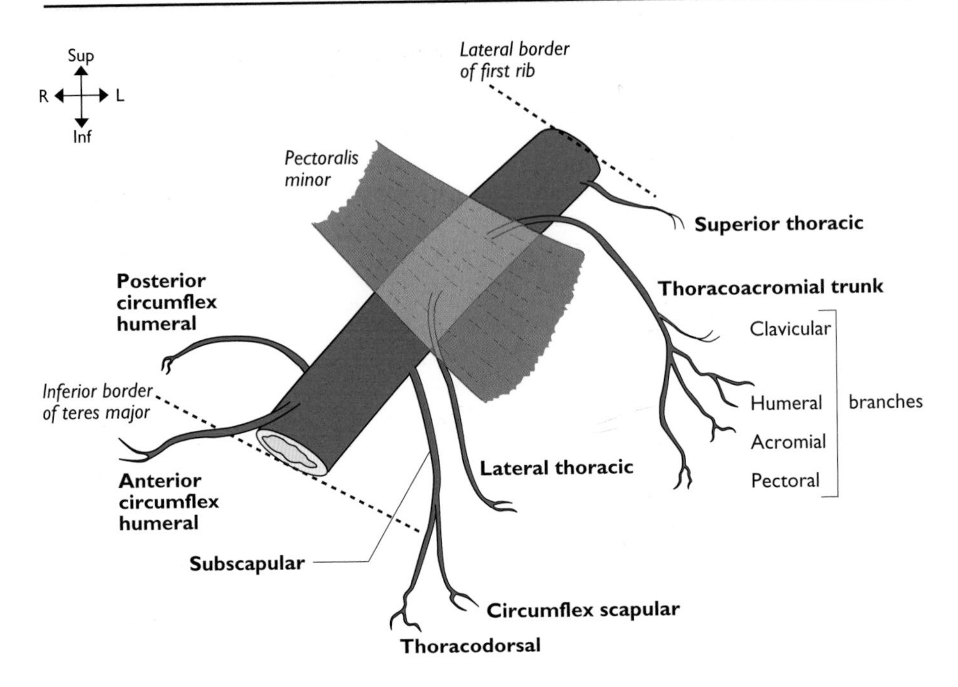

Axillary artery

AXILLARY ARTERY
From: **Subclavian art**
To: **Brachial art**

This is the continuation of the subclavian artery. It commences at the lateral border of the 1st rib and ends at the inferior border of teres major to become the brachial artery. It is divided into three parts by pectoralis minor. It is invested in a fascial sheath arising from the prevertebral fascia.

First part is medial to the upper border of pectoralis minor and has one branch. Anterior to it is the clavipectoral fascia, subclavius and the lateral pectoral nerve. The axillary vein is medial whilst posterior to it are the upper part of serratus anterior, the long thoracic nerve, the medial pectoral nerve and the medial cord of the brachial plexus. Lateral to it are the lateral and posterior cords of the brachial plexus.

Second part has pectoralis minor lying anterior to it and has two branches. Medial to it is the axillary vein and medial cord of the brachial plexus. Posterior to it are the posterior cord and subscapularis whilst lateral to it is the lateral cord of the brachial plexus.

Third part extends from the lower border of pectoralis minor to the inferior border of teres major and has three branches. Anterior to it are pectoralis major, the clavipectoral fascia and the median nerve. Medial to it lie the axillary vein and the ulnar nerve. Posterior to it are the radial nerve, teres major, subscapularis and the tendon of latissimus dorsi. On its lateral side lie the musculocutaneous nerve, lateral root (head) of the median nerve, the tendon of biceps in the bicipital groove and coracobrachialis.

Posterior circumflex humeral artery. Passes posteriorly through the quadrangular space with the axillary nerve to supply shoulder joint and surrounding muscles.

Note. Lateral thoracic artery and pectoral branches of the thoraco-acromial trunk are important supply vessels for the breast.

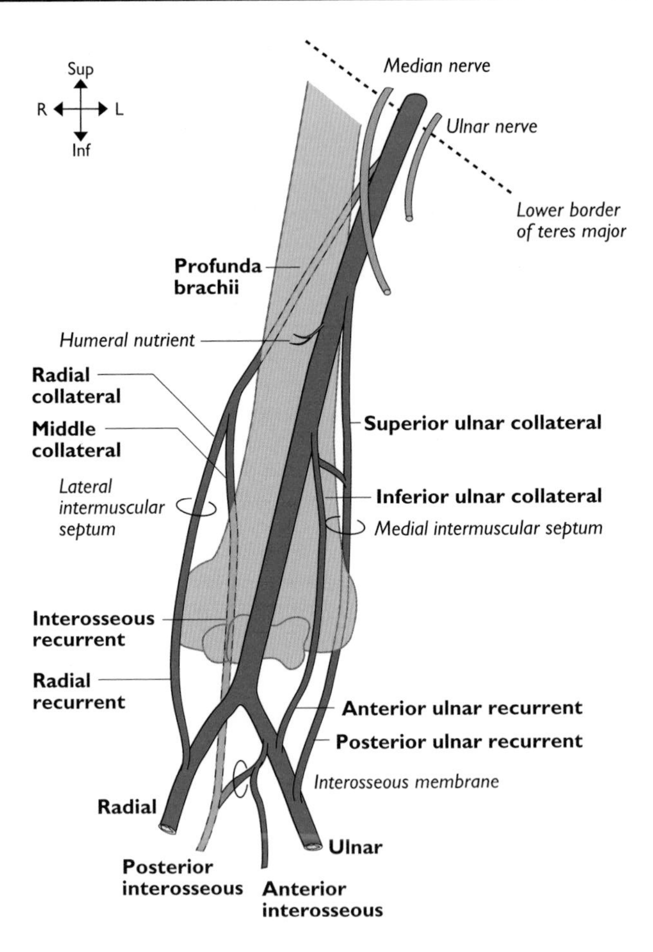

Brachial artery

BRACHIAL ARTERY
From: **Axillary art**
To: **Radial & ulnar arts**

This is the continuation of the axillary artery beginning at the lower margin of the teres major and ending in the cubital fossa at the level of the neck of the radius as the radial and ulnar arteries. At first it lies medial to the humerus then it spirals around to lie anterior to it. It is superficial throughout its course and accompanied by venae commitantes. It is crossed from lateral to medial by the median nerve in the mid arm and by the bicipital aponeurosis in the cubital fossa. Medial to it is the ulnar nerve in the upper arm and, distally, the median nerve. Lateral to it high up are the median and musculo-cutaneous nerves. Coracobrachialis, biceps and its tendon also lie on its lateral side. The artery lies first on the long and then the medial head of triceps, then brachialis in the lower third of the arm.

Arteria profunda brachii. Leaves the posteromedial aspect of the brachial artery just below teres major and passes posteriorly between the long and medial heads of triceps with the radial nerve and into the radial groove before breaking up into its terminal branches.

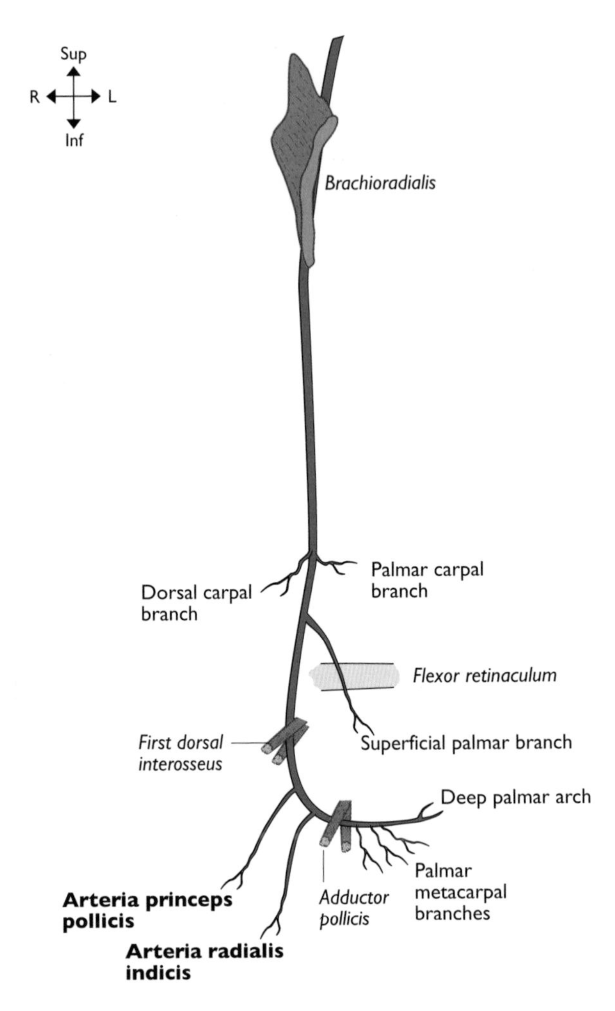

Radial artery

RADIAL ARTERY

From: **Brachial art in midline of cubital fossa**
To: **Deep palmar arch in hand**

The radial artery arises at the terminal bifurcation of the brachial artery in the cubital fossa at the level of the neck of the radius. It crosses anterior to the biceps tendon to lie initially on supinator. It then passes down the radial side of the forearm lying consecutively on pronator teres, the radial head of flexor digitorum superficialis, flexor pollicis longus and the insertion of pronator quadratus before passing onto the lower end of the radius where its pulse is palpable as it lies lateral to the tendon of flexor carpi radialis. It thus lies deep to brachioradialis and, to a lesser extent, flexor carpi radialis. The superficial branch of the radial nerve lies lateral to it in the forearm. It gives off a palmar carpal branch which contributes to the palmar carpal arch. It then gives off a superficial palmar branch (palmar cutaneous branch) which supplies the thenar muscles before anastomosing with the superficial palmar arch. The radial artery then passes beneath the tendons of abductor pollicis longus and extensor pollicis brevis to enter the anatomical snuff box. It passes across the snuff box on the scaphoid and trapezium and under the tendon of extensor pollicis longus. It gives off a dorsal carpal branch to the dorsal carpal arch which in turn supplies the wrist joint, the dorsal aspects of the metacarpals and the dorsal digital arteries. The radial artery then passes down between the two heads of the first dorsal interosseus and, before it enters the palm of the hand, it gives off two named vessels—arteria princeps pollicis (first palmar metacarpal artery) and arteria radialis indicis. The continuation of the radial artery then passes between the two heads of adductor pollicis to become the deep palmar arch which lies 1 cm proximal to the superficial palmar arch (ulnar artery). It supplies the palmar metacarpals, gives off a recurrent branch to the palmar carpal arch and three perforating branches which anastomose with the dorsal metacarpal arteries.

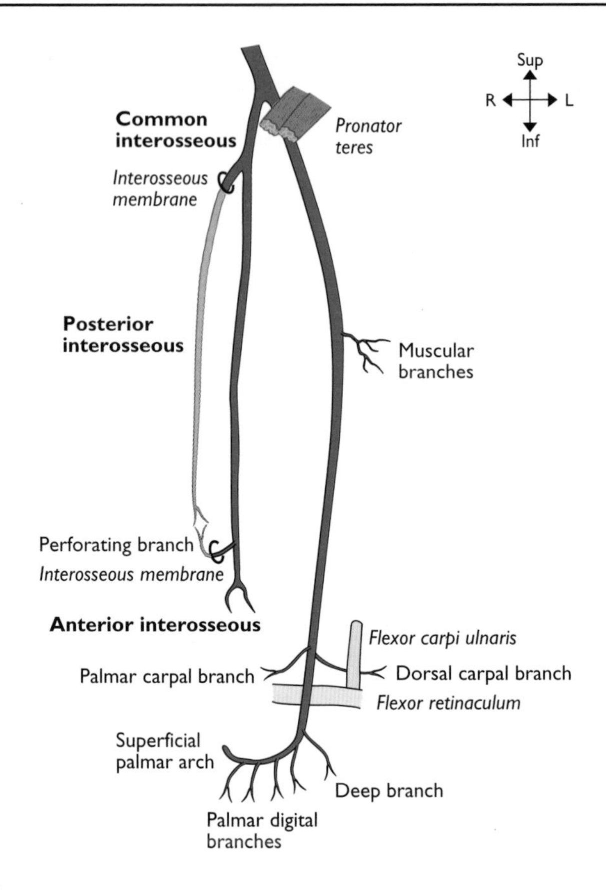

Ulnar artery

ULNAR ARTERY

From: **Brachial art**
To: **Superficial palmar arch in hand**

The artery arises as the terminal bifurcation of the brachial artery in the cubital fossa at the level of the neck of the radius. It leaves the fossa deep to the deep head of pronator teres and deep to the fibrous arch of flexor digitorum superficialis just lateral to the median nerve to cross beneath the nerve before running down the ulnar side of the forearm. It lies on flexor digitorum profundus with the ulnar nerve on its medial side. It lies lateral to flexor carpi ulnaris before passing superficial to the flexor retinaculum. The dorsal and palmar carpal arteries contribute, with similarly named arteries from the radial artery, to the dorsal and palmar carpal arches. The ulnar artery then gives off a deep branch to the deep palmar arch before forming the superficial palmar arch at the level of the distal border of the extended thumb. The superficial arch supplies the hypothenar eminence and gives off the palmar digital arteries. At the level of pronator teres the ulnar artery gives off the common interosseous artery which divides into anterior and posterior interosseous arteries.

Anterior interosseous artery. Descends on the anterior surface of the interosseous membrane together with the anterior interosseous branch of the median nerve lying between flexor digitorum profundus medially and flexor pollicis longus laterally. Branches perforate the membrane to supply the extensor muscles. Above pronator quadratus it gives off a small branch which descends deep to the muscle to join the palmar carpal arch, then the anterior interosseous artery itself passes posteriorly through the membrane to anastomose with the posterior interosseous artery which descends to join the dorsal carpal arch.

Posterior interosseous artery. Passes posteriorly above the interosseous membrane and then runs between supinator superficially and abductor pollicis longus deeply with the deep branch of the radial nerve (posterior interosseous nerve) to descend to supply the extensor muscles of the forearm. It anastomoses with the distal branches of the anterior interosseous artery and dorsal carpal arch.

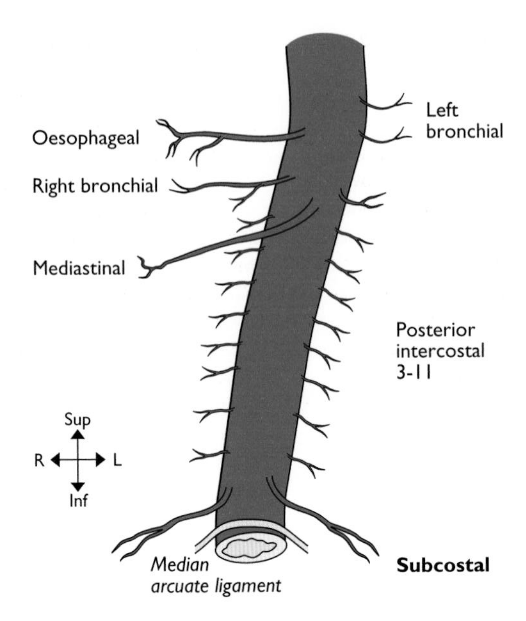

Thoracic (descending) aorta

THORACIC (DESCENDING) AORTA

From: **Arch of aorta**
To: **Abdominal aorta**

This arises as the continuation of the arch of the aorta commencing to the left of the body of T4 and ends as it passes into the abdomen at T12. It grooves the left side of the bodies of T4–T6 vertebrae then it inclines medially to lie in the midline over the lower thoracic vertebrae. It passes out of the thorax at T12 posterior to the median arcuate ligament of the diaphragm to become the abdominal aorta. Lying anterior to it from above down are the hilum of the left lung (particularly the left main bronchus), pericardium, left atrium, oesophagus and diaphragm. Posterior lie the necks of the ribs of T5–T6 and the sympathetic chain at that level, the vertebral bodies and hemiazygos veins. To its right lie the right pleura and lung and thoracic duct. The oesophagus and its surrounding plexus of nerves is initially to its right but lower down it crosses the aorta to lie anterior then slightly to the left. To its left are the left pleura and lung. (Other branch, not illustrated, is pericardial.)

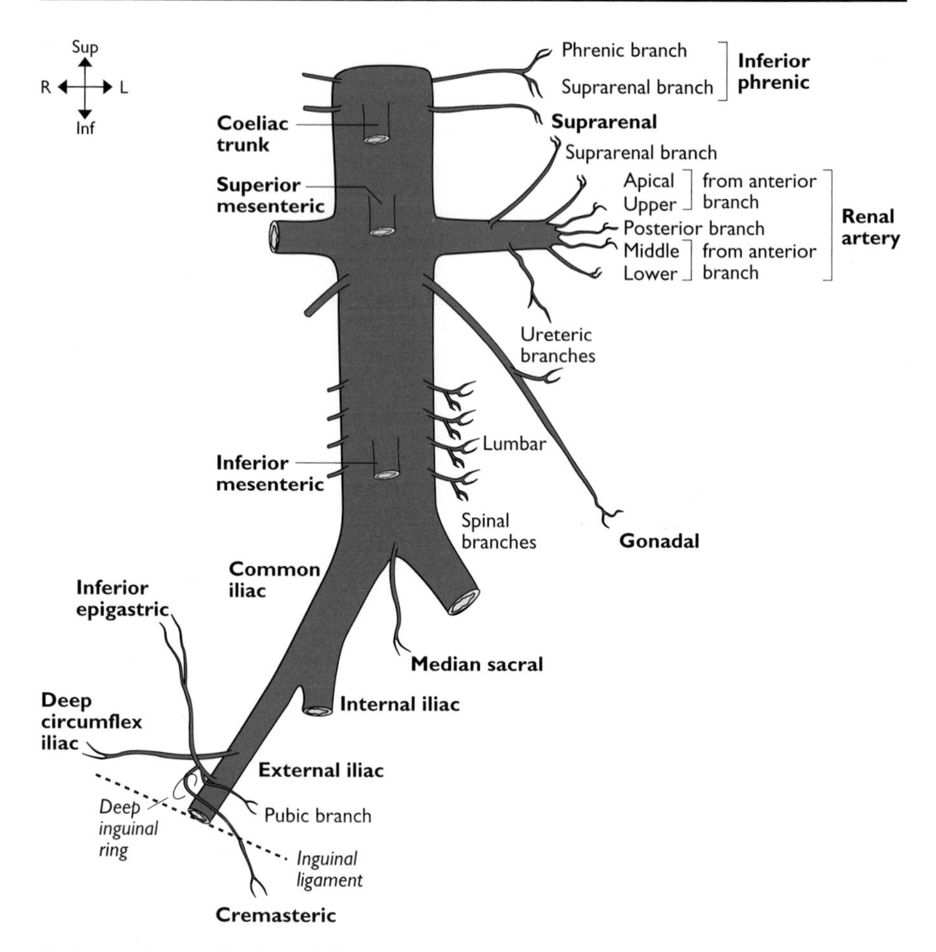

Abdominal aorta & external iliac artery

ABDOMINAL AORTA

From: Thoracic aorta
To: Common iliac arts

This main artery arises as the continuation of the thoracic aorta as it passes, in the midline, posterior to the median arcuate ligament of the diaphragm at T12 and it ends slightly to the left of the midline at L4 where it terminates as the left and right common iliac arteries. Anterior to it, from above downwards, are the coeliac trunk and its branches, the coeliac plexus, lesser sac, superior mesenteric artery, left renal vein, body of pancreas, commencement of each gonadal artery, fourth part of the duodenum, posterior parietal peritoneum, attachment of the mesentery and inferior mesenteric artery. Posterior to it are the lumbar arteries and left lumbar veins, anterior longitudinal ligament and vertebral bodies with their intervertebral discs. To its right are the cisterna chyli, thoracic duct, azygos vein, right crus of diaphragm and inferior vena cava. To its left are the left crus of diaphragm, left coeliac ganglion, the duodenojejunal flexure (upper border of L2), sympathetic trunk and inferior mesenteric vessels. On both sides the phrenic, suprarenal and renal vessels are lateral whilst distal to the bifurcation of the abdominal aorta is the median sacral artery.

Common iliac arteries. These commence at the bifurcation of the abdominal aorta just to the left of the midline at L4 and pass inferolaterally to the level between L5 and S1 vertebrae where they bifurcate anterior to the sacro-iliac joint to give the external and internal iliac arteries. Anterior to each vessel are sympathetic contributions to the superior hypogastric plexus, the ureter (near the terminal bifurcation of the vessel), peritoneum and small bowel. In addition on the left side the superior rectal artery lies anterior. Posterior to each vessel are the sympathetic trunk, obturator nerve, lumbosacral trunk, iliolumbar artery and the bodies of L4 and L5 with the disc between them. In addition posteriorly on the right side are the terminal portions of the common iliac veins and the commencement of the inferior vena cava. The left common iliac vein lies posteromedial to the left common iliac artery. Psoas major lies lateral to each vessel.

Gonadal artery. Descends passing obliquely inferiorly on the posterior abdominal wall to the level of the external iliac artery. The testicular arteries pass around the lower border of the false pelvis to enter the inguinal canal through the deep ring to form part of the spermatic cord. The ovarian vessels descend over the external iliac vessels into the infundibulopelvic fold to supply the ovary via the broad ligament. The common relations of the arteries in both sexes are: left, posterior—psoas, genitofemoral nerve, ureter and external iliac artery; left, anterior—inferior mesenteric vein, left colic artery and sigmoid mesentery; right, posterior—inferior vena cava, psoas, genitofemoral nerve, ureter and external iliac artery; right, anterior—third part of the duodenum, right colic artery and ileal mesentery.

EXTERNAL ILIAC ARTERY

From: Common iliac art
To: Femoral art

The external iliac artery descends laterally from the common iliac artery to pass under the inguinal ligament at the mid inguinal point (half way between the anterior superior iliac spine and symphysis pubis) where it becomes the femoral artery.

Posterior and lateral to it is the medial border of psoas major whilst the femoral vein comes to lie medially. Anteromedially it is covered by peritoneum on which lies small bowel with sigmoid colon additionally on the left. It is crossed at its origin by the ureter and then by the gonadal vessels, genital branch of the genitofemoral nerve, deep circumflex iliac vein and vas deferens or round ligament.

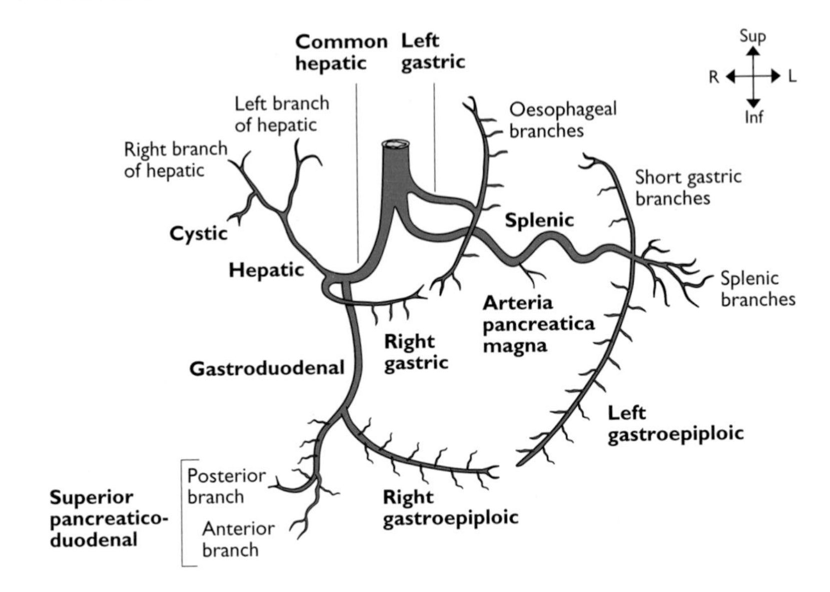

Coeliac trunk

COELIAC TRUNK

From: **Abdominal aorta at lower border of T12**

To: **Terminal brs—left gastric, splenic & common hepatic arts**

The coeliac trunk (axis) arises from the anterior aspect of the abdominal aorta at the level of the lower border of T12 and after 1 cm divides into its three terminal branches.

Left gastric artery. Passes superolaterally on the posterior wall of the lesser sac to reach the apex of this structure at the cardio-oesophageal junction where it divides into oesophageal branches to supply the lower third of the oesophagus through the oesophageal opening in the diaphragm. Its terminal gastric branches run inferiorly along the upper portion of the lesser curve of the stomach to anastomose with the right gastric artery.

Splenic artery. Passes laterally to the left, angled slightly superiorly, running in the posterior wall of the lesser sac. Its course is markedly tortuous as it runs along the superior border of the pancreas, passing anterior to the left crus of the diaphragm, the upper pole of the left kidney and the left suprarenal gland before entering the lienorenal ligament to reach the splenic hilum.

Common hepatic artery. Runs inferolaterally to the right in the posterior wall of the lesser sac towards the first part of the duodenum where it gives off first the gastroduodenal and then right gastric arteries. It then curves anteriorly as the hepatic artery to pass into the peritoneal reflection which forms the inferior margin of the opening of the lesser sac. It approaches the portal vein from its left side and then comes to lie anterior to it, with the bile duct on its right, as it ascends in the free border of the lesser omentum (anterior margin of the foramen of Winslow or epiploic foramen) before terminating at the porta hepatis as the right and left hepatic branches.

Gastroduodenal artery. Descends directly behind the first part of the duodenum to the left of the bile duct and divides at the upper border of the pancreas into terminal branches.

Right gastric artery. Arises from the hepatic artery as it enters the lesser omentum and passes along the lesser curve of the stomach to anastomose with the left gastric artery.

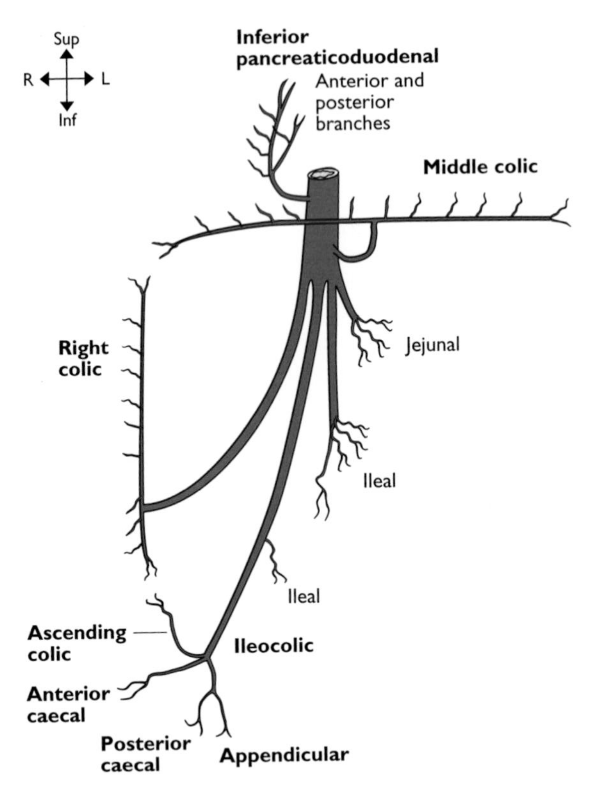

Superior mesenteric artery

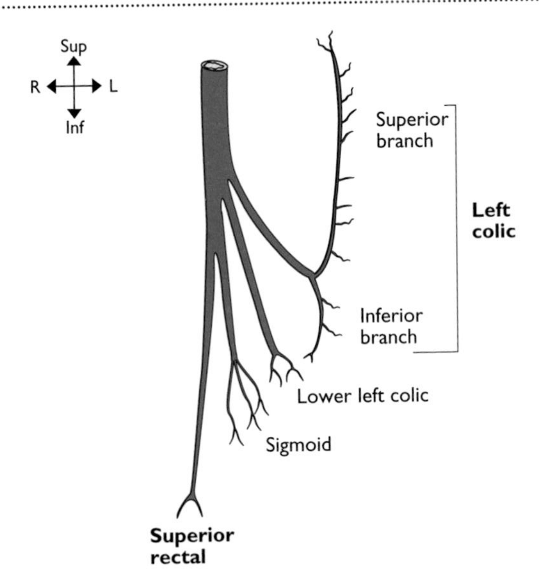

Inferior mesenteric artery

SUPERIOR MESENTERIC ARTERY

From: **Abdominal aorta**
To: **Terminal brs**

The superior mesenteric artery arises from the anterior surface of the abdominal aorta at the level of L1. It passes inferiorly over the left renal vein with the splenic vein and body of pancreas anterior to it. It next lies on the uncinate process of the pancreas and the junction of the third and fourth parts of the duodenum from where it passes obliquely and to the right into the mesentery of the small bowel before giving off its terminal branches. The superior mesenteric vein is on its right whilst posterior to the terminal branches are the inferior vena cava, the right ureter and psoas major. The superior mesenteric plexus of nerves surrounds the artery. It supplies bowel from the mid second part of the duodenum, jejunum, ileum, ascending and right two-thirds of transverse colon.

Inferior pancreaticoduodenal artery. Leaves the superior mesenteric artery as it begins to cross the duodenum and divides into an anterior and posterior branch. The anterior branch passes to the right to anastomose with the anterior superior pancreatico-duodenal artery anterior to the head of the pancreas. The posterior branch passes also to the right but posterior to the head of the pancreas to anastomose with the posterior superior pancreaticoduodenal artery.

Middle colic artery. Note that this artery arises early from the superior mesenteric artery to supply the transverse colon. This is logical as the right colic would otherwise need to be excessively long.

Ileocolic artery. Passes obliquely inferiorly to the right in the root of the mesentery where it passes anterior to the right ureter and right gonadal vessels to reach the caecum where it divides into its terminal branches.

INFERIOR MESENTERIC ARTERY

From: **Abdominal aorta**
To: **Terminal brs**

This artery arises from the anterior surface of the abdominal aorta at the level of L3 posterior to the third and fourth part of the duodenum. It passes inferiorly and to the left crossing the left common iliac artery medial to the left ureter. The inferior mesenteric vein lies on its left (lateral) side. It divides into its terminal branches in the descending mesocolon. It supplies the left third of the transverse colon, the descending and sigmoid colon and the rectum to the dentate line of the anus.

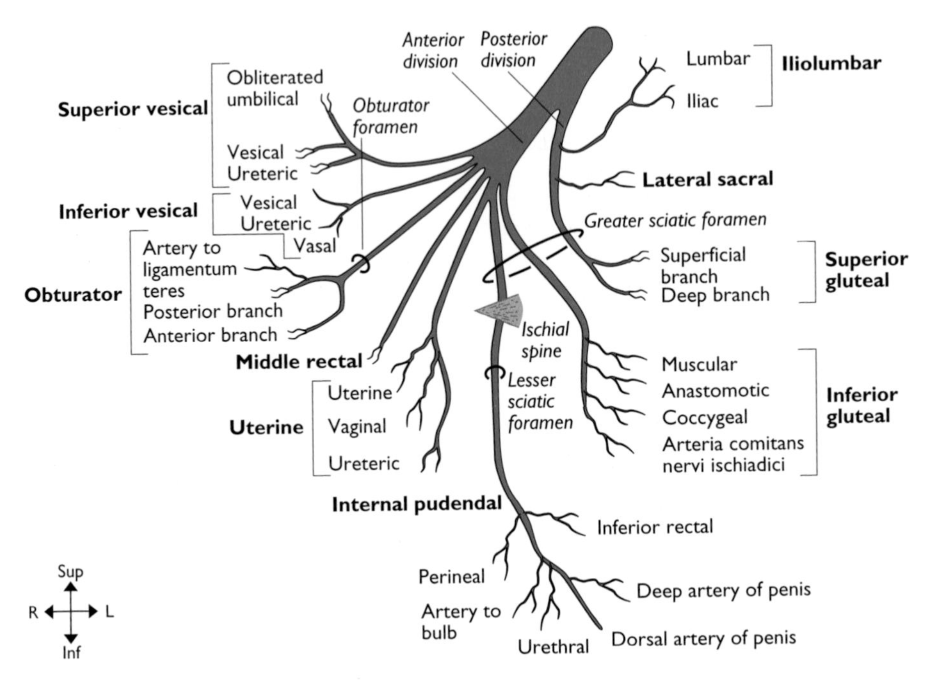

Internal iliac artery

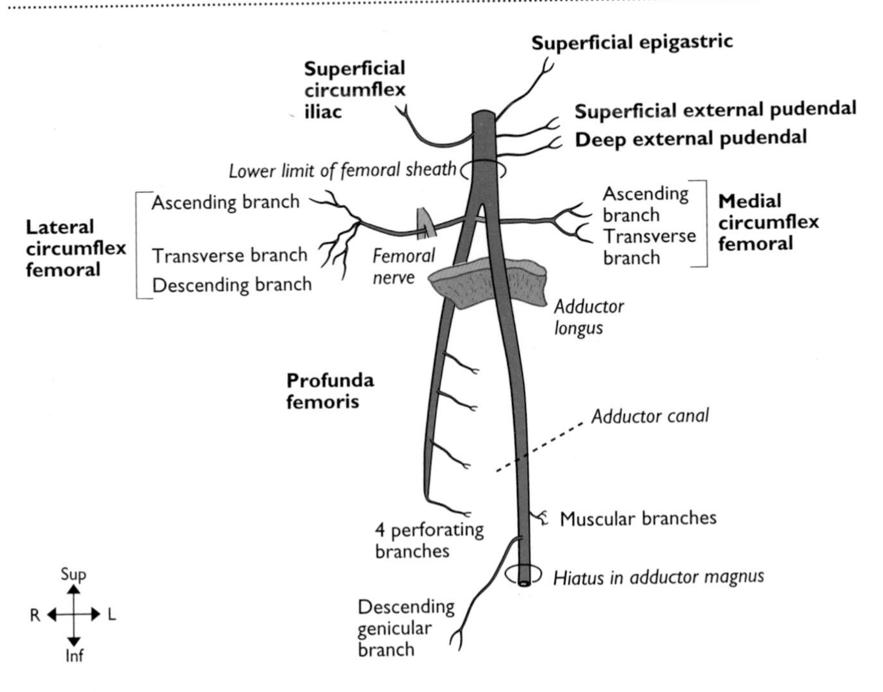

Femoral artery

INTERNAL ILIAC ARTERY

From: **Common iliac art**
To: **Terminal brs**

The artery commences at the level of the disc between L5 and S1 and passes posteriorly into the pelvis for 4 cm before forming anterior and posterior divisions which break up into their terminal branches. Anterior lie the ureter and fallopian tube and ovary in the female. Posterior are the internal iliac vein, lumbosacral trunk and sacro-iliac joint. On the lateral side are the external iliac artery and vein, obturator nerve and psoas major. The parietal peritoneum and small bowel lie medially.

Internal pudendal artery. Arises from the anterior division of the internal iliac artery and descends on the lateral wall of the pelvis towards the greater sciatic foramen. It leaves the pelvis via this foramen, inferior to piriformis, before passing over the tip of the ischial spine to enter the ischio-anal fossa via the lesser sciatic foramen. In runs on the lateral wall of the ischio-anal fossa on obturator internus in the pudendal (Alcock's) canal. It passes into the deep perineal pouch where it gives off its terminal branches. (Other branches of the perineal branch, not illustrated, are transverse perineal and posterior scrotal.)

Note: In the female the vaginal artery is equivalent to the inferior vesical artery in the male, and the uterine artery is equivalent to the middle rectal artery. The round ligament is supplied by the uterine artery whilst the vas deferens is usually supplied by the inferior vesical or less often by the superior vesical artery.

FEMORAL ARTERY

From: **External iliac art**
To: **Popliteal art**

This is the continuation of the external iliac artery and commences posterior to the inguinal ligament at the mid-inguinal point (half way between the anterior superior iliac spine and the symphysis pubis). It ends as it passes through the adductor hiatus in adductor magnus to become the popliteal artery. It emerges from under the inguinal ligament with the femoral vein medial to it, both within the femoral sheath. Lateral to it and outside the femoral sheath is the femoral nerve. It lies on the tendon of psoas major and is separated from pectineus and adductor longus by the femoral vein which comes to lie progressively more posterior to the artery within the femoral triangle. As the femoral artery enters the adductor canal it lies on adductor longus then adductor magnus. It is covered initially only by deep fascia then by sartorius; the saphenous nerve passes anteriorly from lateral to medial. Anterolateral to the artery is vastus medialis.

Profunda femoris is the main branch of the femoral artery which is given off posterolaterally just below the femoral sheath 3.5 cm below the inguinal ligament. It runs posteriorly between pectineus and adductor longus to pass into the deep thigh where it provides the deep structures and the posterior and medial compartments with their main arterial supply. Perforating and descending branches anastomose with the genicular branches of the popliteal artery.

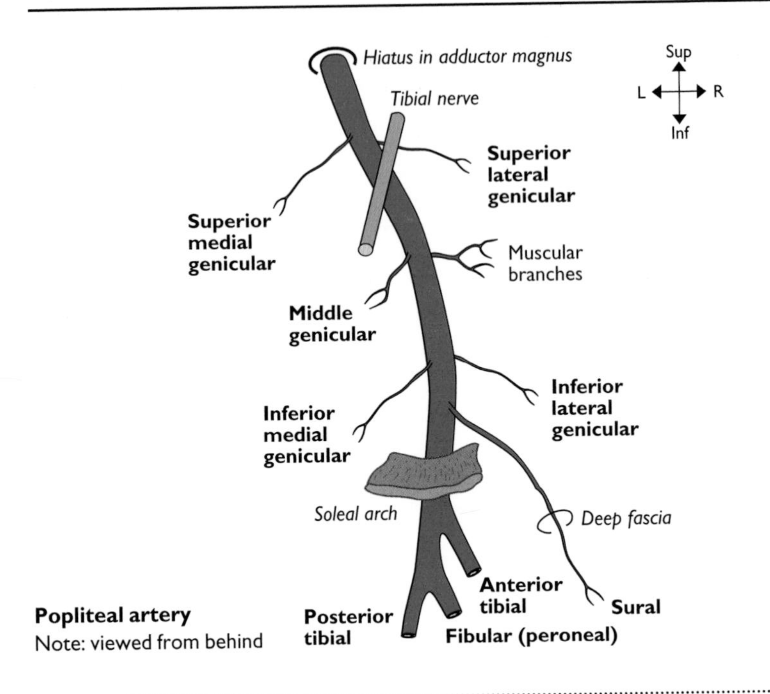

Popliteal artery
Note: viewed from behind

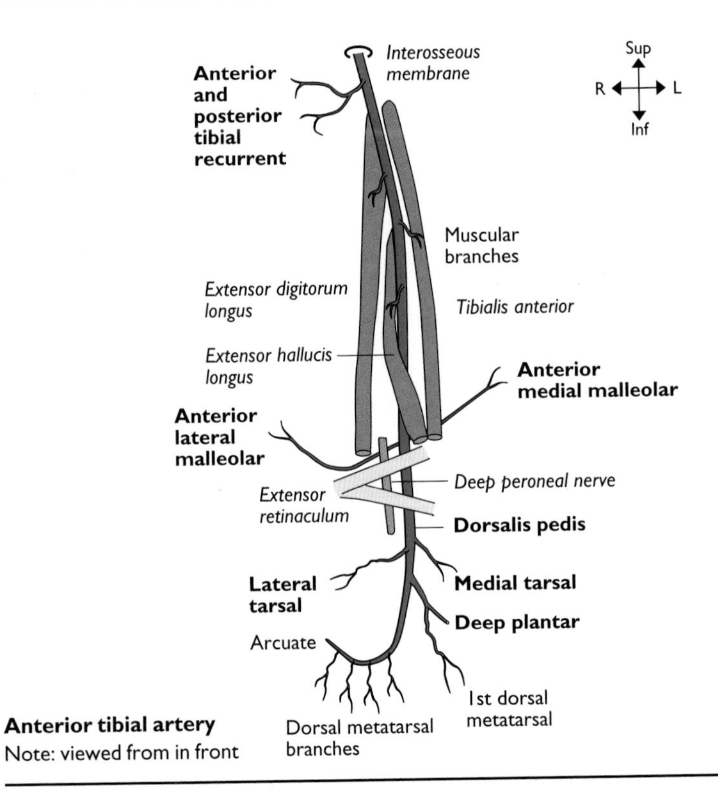

Anterior tibial artery
Note: viewed from in front

POPLITEAL ARTERY
From: Femoral art
To: Ant & post tibial arts

This artery commences as the continuation of the femoral artery as the latter passes through the hiatus in adductor magnus and ends as it passes under the fibrous arch of soleus where it immediately divides into anterior and posterior tibial arteries. The popliteal artery extends from a hand's breadth above the knee and to the same distance below it. It enters the popliteal fossa medial to the femur and becomes the deepest structure, lying with only fat between it and the popliteal surface of the femur. Lower down it lies on the capsule of the knee joint and then on popliteus. Biceps femoris is lateral to it and semimembranosus medial. Lower down it lies between the two heads of gastrocnemius. It is crossed laterally to medially by the tibial nerve and the popliteal vein with the vein always between the artery and nerve.

ANTERIOR TIBIAL ARTERY
From: Popliteal art
To: Dorsalis pedis art

This artery commences at the bifurcation of the popliteal artery just under the fibrous arch of soleus, at the distal border of popliteus. It supplies the structures in the extensor compartment of the lower leg. It passes anteriorly between the heads of tibialis posterior to pass above the upper border of the interosseous membrane, medial to the neck of the fibula accompanied by its venae commitantes. It descends on the interosseous membrane and crosses the lower tibia at the ankle joint, mid way between the malleoli and there becomes the dorsalis pedis artery. Initially it lies between tibialis anterior (medially) and extensor digitorum longus (laterally), then between tibialis anterior and extensor hallucis longus. At the ankle it is crossed anteriorly by the extensor retinacula and also from lateral to medial by the tendon of extensor hallucis longus. The deep fibular (peroneal) nerve is initially lateral to the artery high up in the extensor compartment but passes anterior to it half way down the leg, becoming lateral to it again under the extensor retinaculum. The anterior tibial veins run in close association with the artery throughout.

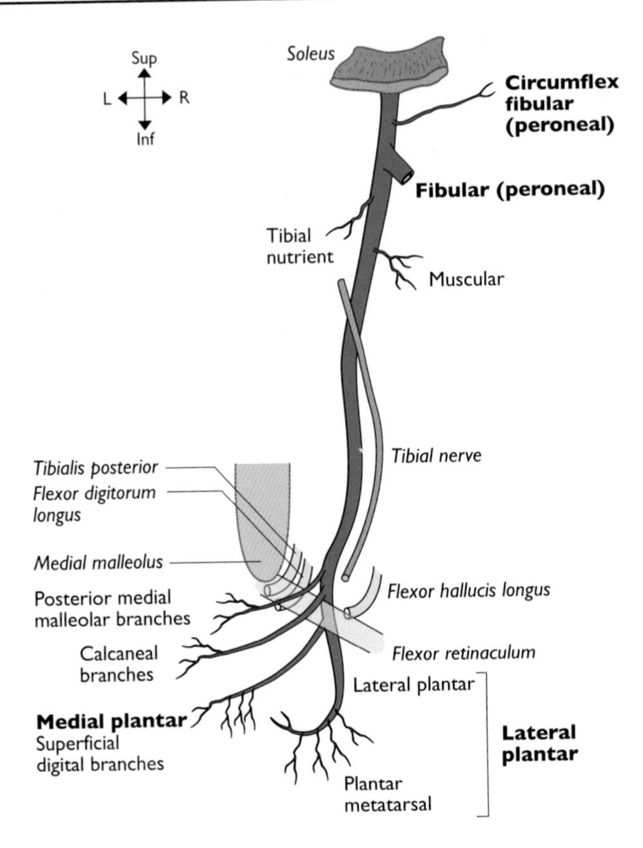

Posterior tibial artery
Note: viewed from behind

POSTERIOR TIBIAL ARTERY
From: Popliteal art
To: Med & lat plantar arts

This artery arises at the bifurcation of the popliteal artery just under the fibrous arch of soleus, at the lower border of popliteus, and ends by bifurcating into the medial and lateral plantar arteries deep to abductor hallucis. It supplies structures in the posterior compartment of the lower leg. It is accompanied by venae commitantes and lies from above downwards on tibialis posterior, flexor digitorum longus, the tibia and the ankle joint. It lies deep to gastrocnemius, soleus, the flexor retinaculum and abductor hallucis. Posterior to the medial malleolus it lies between the tendon of flexor digitorum longus and the tibial nerve which crosses posterior to the artery mid way down the calf from the medial side to become postero-lateral. (Other branch, not illustrated, is a communicating branch to the fibular (peroneal) artery.)

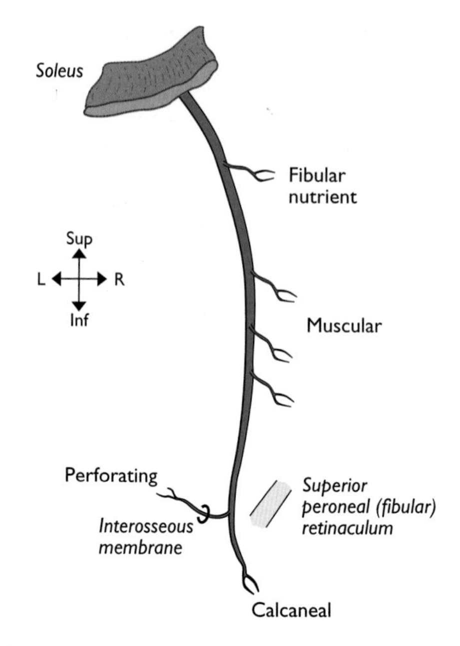

***Fibular (peroneal)* artery**

FIBULAR (PERONEAL) ARTERY

From: Post tibial art
To: Terminal brs

This is a branch of the posterior tibial artery arising 2.5 cm below its origin under soleus. It supplies structures in the lateral compartment of the lower leg. Due to the proximity of origin of the three terminal branches of the popliteal artery this point of origin is commonly referred to as the 'popliteal trifurcation'. It passes inferolaterally to reach and run along the medial crest of the fibula between tibialis posterior and flexor hallucis longus to divide into its terminal branches at the level of the inferior tibiofibular joint and the superior fibular (peroneal) retinaculum. Thus, although it supplies the fibular (peroneal) compartment by branches that pass laterally, the main fibular (peroneal) artery itself remains in the posterior compartment. Above, it is covered by soleus and deep fascia whilst in the lower leg flexor hallucis longus crosses it from lateral to medial.

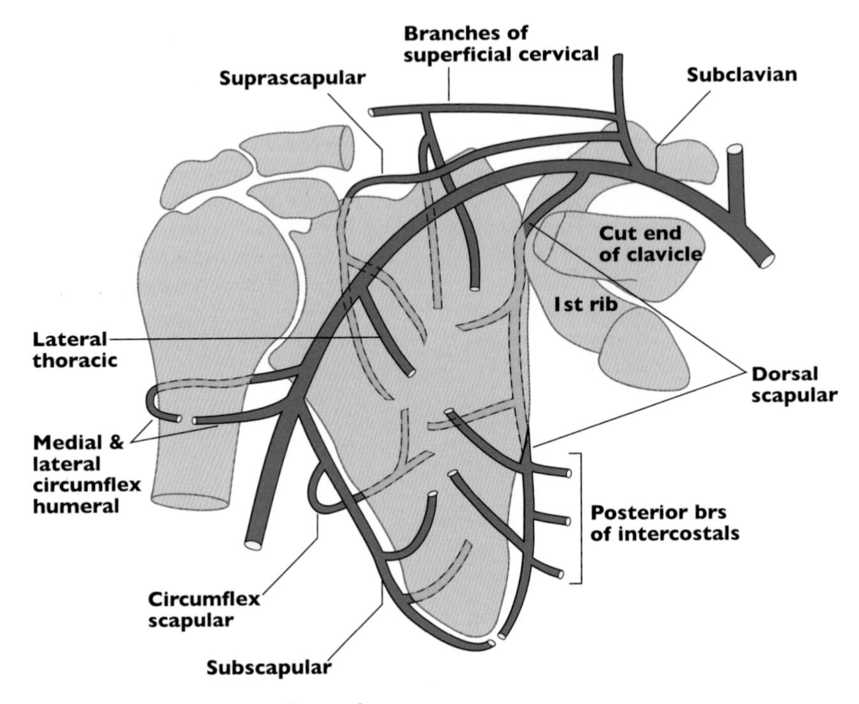

Arterial anastomoses around scapula

Note: The superficial cervical artery is named "transverse cervical artery" if it gives origin to the dorsal scapular artery instead of the latter arising separately from the second part of the subclavian artery

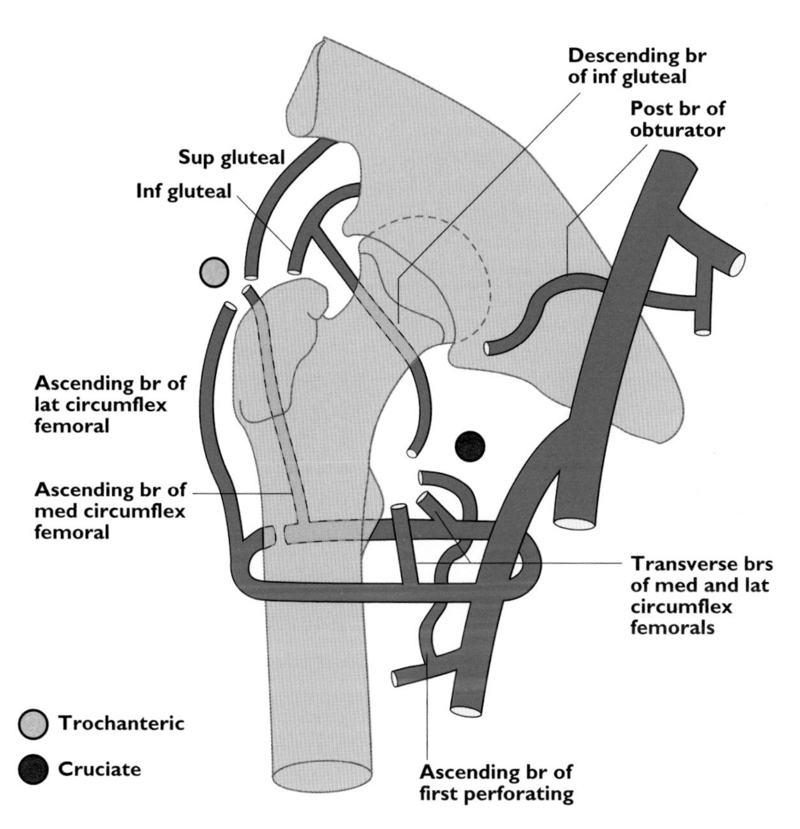

Arterial anastomoses around hip

2: VEINS

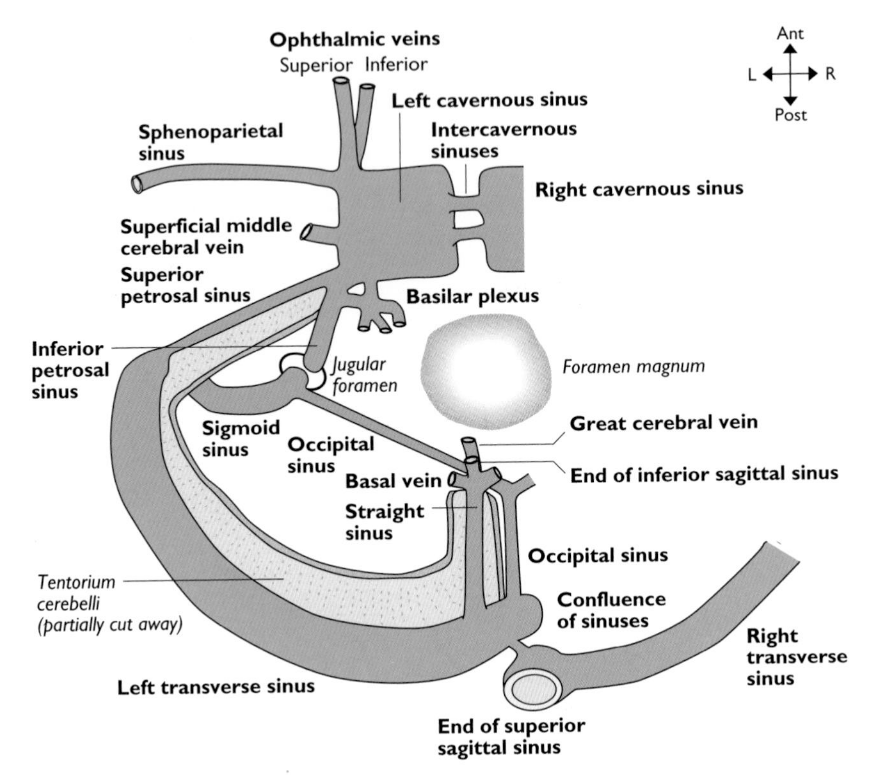

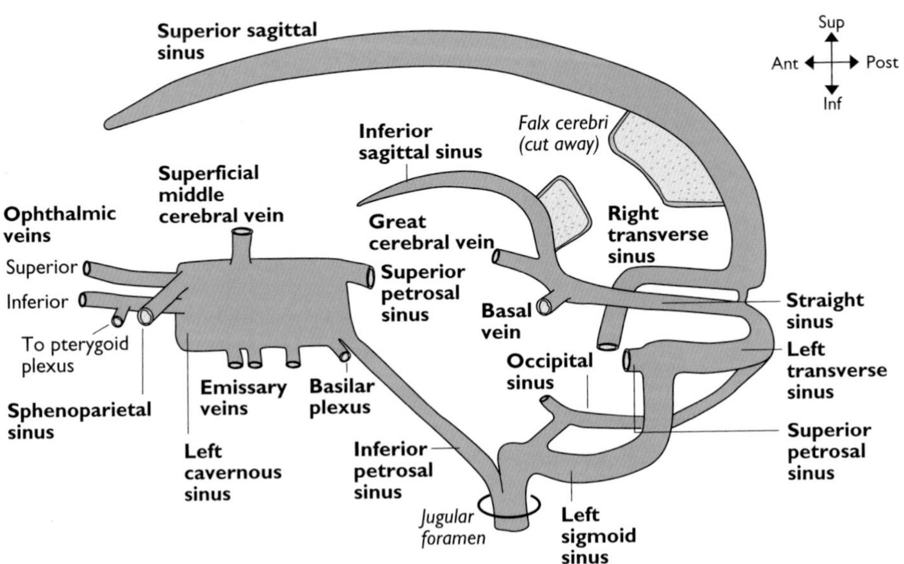

Intracranial sinuses & veins

INTRACRANIAL SINUSES & VEINS

From: **Cerebrum, cerebellum & diploe of the skull**
To: **Internal jugular V**

The cerebrum, cerebellum and bones of the skull are drained by the external, internal and meningeal veins to the sinuses. The sinuses lie between the endosteal and meningeal layers of the dura mater, either as an endothelial lined space in its free edge (inferior sagittal and straight sinuses), or a similarly lined space where the dura is reflected over the bone of the inner surface of the skull. They are characteristically thin walled, contain no valves and communicate freely with each other.

Superior sagittal sinus. Lies in the superior margin of the falx cerebri draining the arachnoid granulations as it does so. It commences at the foramen caecum and, posteriorly, usually drains as a continuation into the right transverse sinus. It frequently connects at its termination with the left transverse sinus.

Inferior sagittal sinus. Runs in the inferior free margin of the falx cerebri draining medial cortical veins as it does so, and terminates by fusing with the great cerebral vein of Galen and right and left basal veins to form the straight sinus.

Straight sinus. Runs in the junction of the falx cerebri and tentorium cerebelli for a short distance before terminating in its continuation—the left transverse sinus.

Transverse (lateral) sinus. Runs in the lateral border of the tentorium cerebelli grooving the occipital and squamous temporal bones, to terminate in the sigmoid sinus just as it receives the superior petrosal sinus from the cavernous sinus on each side.

Sigmoid sinus. Deeply grooves the temporal bone as it passes inferomedially into the posterior compartment of the jugular foramen at the inferior margin of which it unites with the inferior petrosal sinus to form the internal jugular vein.

Cavernous sinus. Lies on the lateral wall of the body of the sphenoid bone and is a lateral relation of the sella turcica, the pituitary gland and the sphenoidal air sinus. It lies medial to the medial gyrus of the temporal lobe. Lying in it is the internal carotid artery (carotid syphon) with the abducent nerve (VI) on its lateral surface and lying on its lateral wall are nerves (from above down): oculomotor (III), trochlear (IV), ophthalmic (Va) and maxillary (Vb) divisions of the trigeminal. It has a sponge-like reticular structure and its connections, particularly those with the other major sinuses (as shown opposite), frequently provide both supply to, and drainage from, the sinus. There are two intercavernous sinuses connecting the cavernous sinuses to each other.

Occipital sinus. Begins at the foramen magnum and ascends to end in the confluence of sinuses.

Confluence of sinuses is at the lowest, posterior end of the superior sagittal sinus at the point that it turns, usually to the right, to become the transverse sinus. It connects with the straight, occipital and opposite transverse sinuses.

Sphenoparietal sinus. Runs along the lesser wing of the sphenoid bone and drains into the cavernous sinus.

Superior petrosal sinus. Runs along the petrous temporal bone where the edge of the tentorium cerebelli attaches and hence connects the cavernous and transverse sinuses.

Inferior petrosal sinus. Runs inferiorly to connect the cavernous sinus to the internal jugular vein. It exits the skull through the anterior compartment of the jugular foramen with the glossopharyngeal nerve.

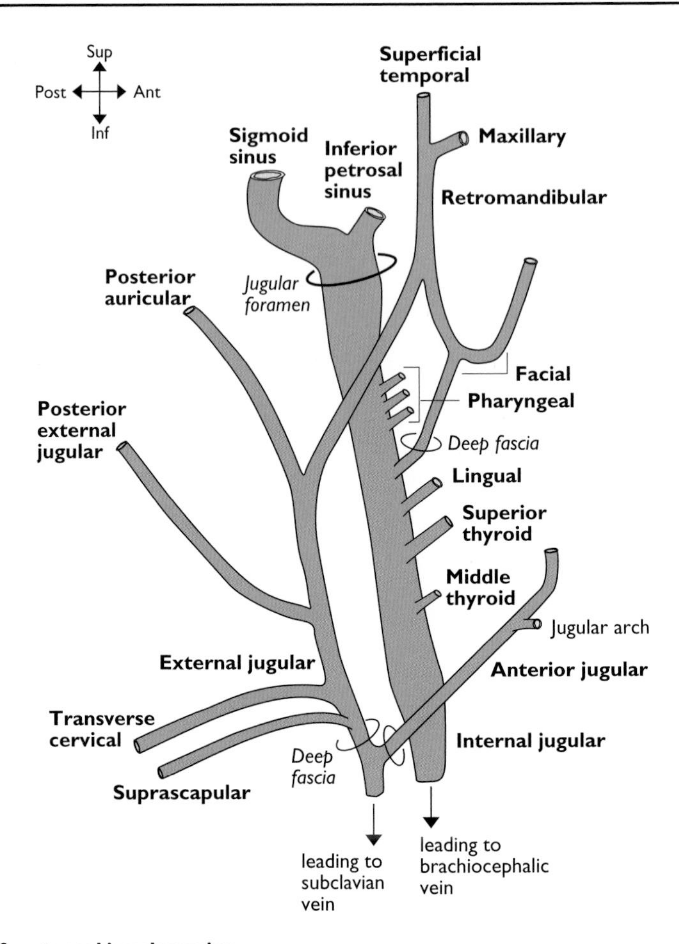

Internal & external jugular veins

INTERNAL JUGULAR VEIN
From: **Sigmoid & inf petrosal sinuses**
To: **Brachiocephalic Vs**

It runs almost vertically downwards within the carotid sheath although its covering is thin and readily stretched. Its relationship to the internal carotid artery is as follows: posterior to the artery at the level of C2, posterolateral at C3 and lateral at C4, the vagus nerve (X) lies between the two throughout. Outside the sheath it is surrounded by deep cervical lymph nodes and it lies on (from above down): the lateral mass of the atlas (C1), prevertebral fascia, scalenus medius, scalenus anterior and the dome of the cervical pleura. It is crossed at its origin by the spinal root of the accessory nerve, the lower root of the ansa cervicalis in its middle third and is overlaid by sternocleidomastoid and the tendon of omohyoid in its lower third.

EXTERNAL JUGULAR VEIN
From: **Various brs**
To: **Subclavian V**

It arises from the junction of the posterior auricular vein and the posterior division of the retromandibular vein and drains into the subclavian vein. The external jugular system lies within the superficial tissues of the neck (as does the anterior jugular system). The external and anterior jugular veins pierce the deep fascia of the neck, usually posterior to the clavicular head of sternocleidomastoid to fuse before draining into the subclavian vein.

2

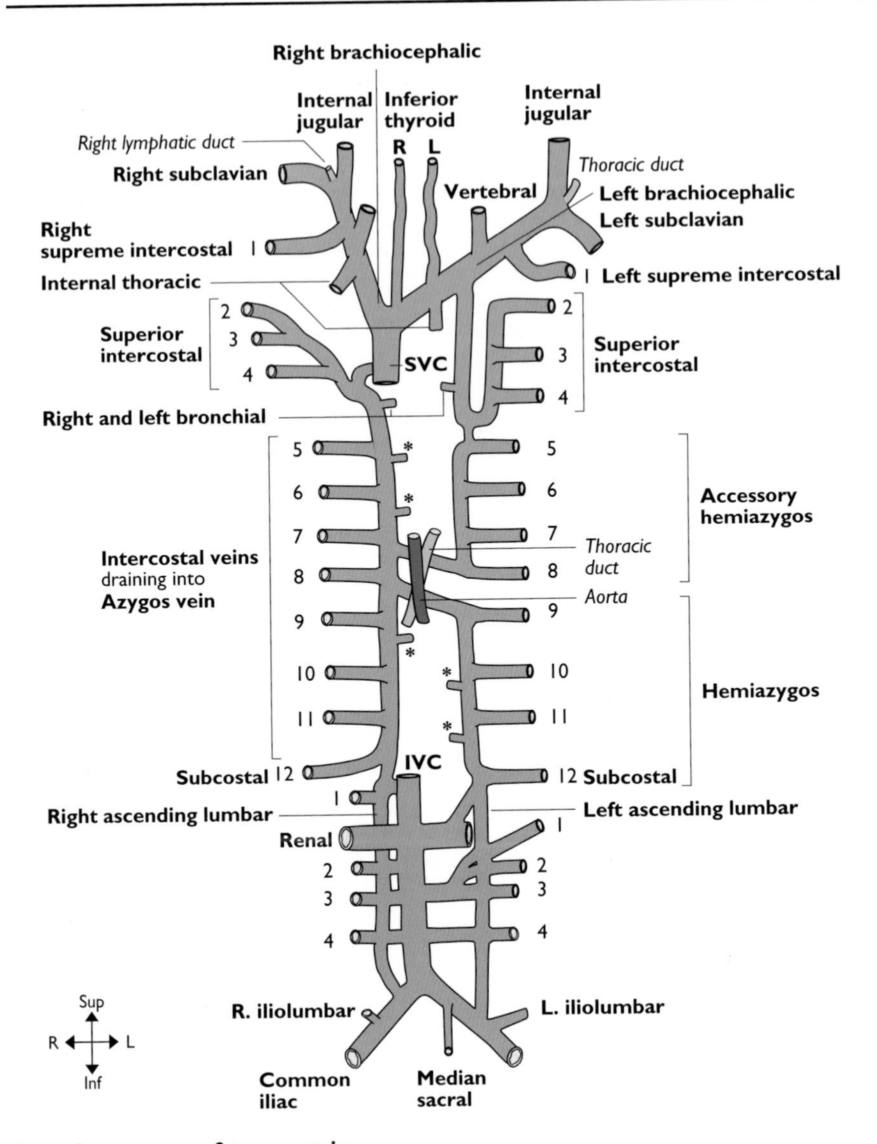

Superior vena cava & azygos veins

Note: (1) Accessory hemiazygos crosses to azygos at T7 and hemiazygos at T8—each crosses behind thoracic aorta, oesophagus and thoracic duct; (2) left bronchial vein may enter accessory hemiazygos. *= oesophageal and mediastinal veins, IVC = inferior vena cava, SVC = superior vena cava

SUPERIOR VENA CAVA

From: **Brachiocephalic Vs**
To: **Right atrium**

It is formed posterior to the right first costal cartilage and passes posterior to the right sternal border where it is a close posterior relation of the right internal thoracic vessels and sternal periosteum and is occasionally overlaid by the anterior segment of the right upper lobe of the lung. It lies anterolateral to the trachea and upper right lung hilum with the right phrenic nerve applied to its right lateral surface. It receives the azygos vein into its posterior surface at the level of T4. It enters the superior surface of the right atrium without any valvular mechanism guarding its orifice.

Left brachiocephalic vein. Formed posterior to the left sternoclavicular joint and anterior to the cervical pleura. It passes obliquely downwards to the right, posterior to the manubrium, separated from it only by the thymus gland or its remnant. It lies anterior to the left common carotid and brachiocephalic arteries and crosses the upper anterior aortic arch. Unlike the right superior intercostal vein (see below) the left superior intercostal vein drains into the left brachiocephalic vein. Other tributaries (not shown) are thymic and pericardial veins.

Right brachiocephalic vein. Formed posterior to the right sternoclavicular joint and passes directly inferiorly behind the right side of the manubrium, anterolateral to the trachea and anteromedial to the pleura over the upper lobe of the lung.

AZYGOS VEINS

From: **Inf vena cava/ascending lumbar Vs**
To: **Sup vena cava**

The azygos veins drain the upper lumbar region and the thoracic wall. There is a single system on the right whilst on the left there are two—the hemiazygos and accessory hemiazygos that drain over into the azygos separately.

Azygos vein. Arises at the approximate level of the right renal vein either as a posterior tributary of the inferior vena cava or as a confluence of the right ascending lumbar and right subcostal vein. It passes through the aortic opening of the diaphragm under the right crus at the level of T12 vertebra and ascends on the right side of the vertebral bodies behind the oesophagus. It turns anteriorly to pass over the hilum of the right lung, lateral to the oesophagus, trachea and right vagus, to enter the superior vena cava at the level of T4. Its tributaries are the lower eight right posterior intercostal veins, the right superior intercostal vein (draining the 2nd, 3rd and 4th right intercostal veins), bronchial and oesophageal veins and, from the left side, the two hemiazygos veins.

Hemiazygos vein. Arises from the confluence of the left ascending lumbar vein, the left subcostal vein and often a tributary from the left renal vein. It ascends through the aortic opening of the diaphragm and onto the left side of the thoracic vertebra to the level of T9 from where it crosses posterior to the aorta, oesophagus and thoracic duct to enter the azygos vein at T8. It drains the four lower left posterior intercostal veins (9–12th).

Accessory hemiazygos vein. Drains the 5–8th left posterior intercostal veins and runs inferiorly on the left side of the vertebral bodies to T8 where it crosses similarly to the hemiazygos vein to enter the azygos vein at T7. It also receives tributaries from the bronchial and mid-oesophageal veins.

Note: The anterior intercostal veins drain to the musculophrenic and internal thoracic veins.

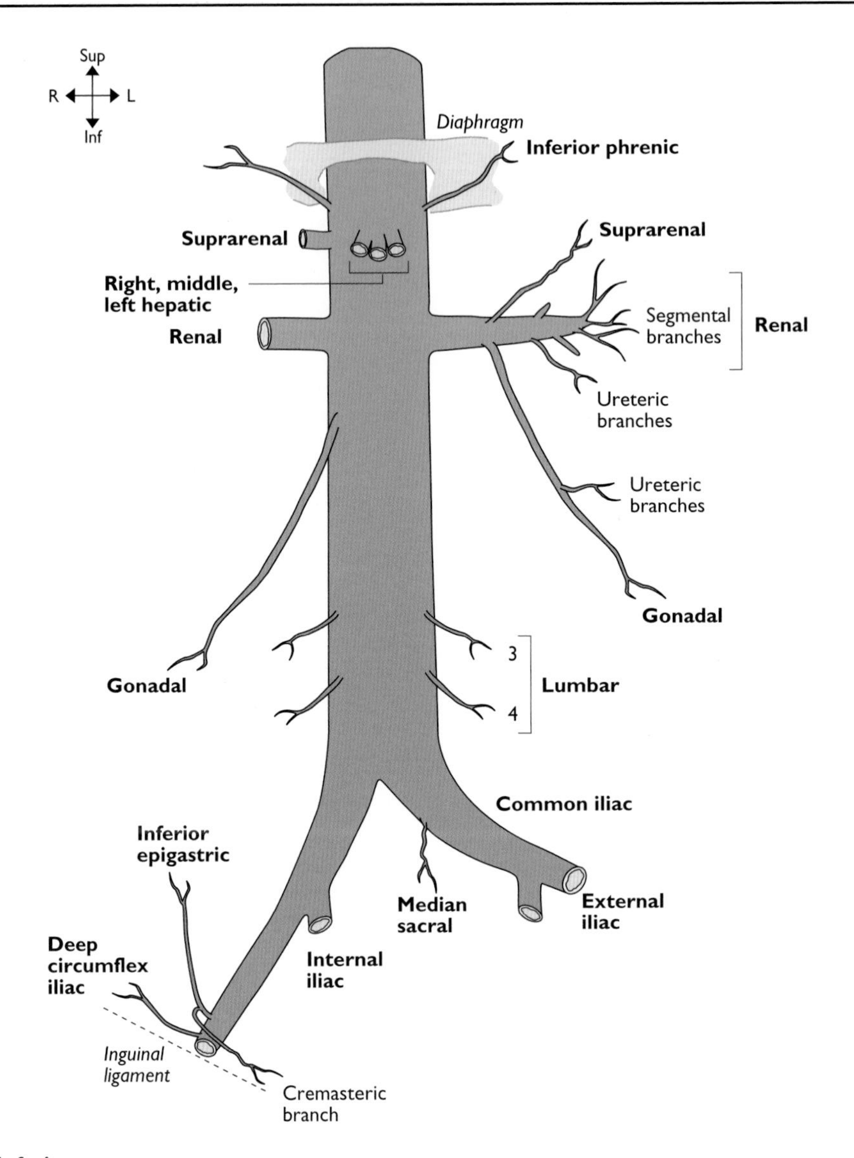

Inferior vena cava

INFERIOR VENA CAVA
From: **Common iliac Vs**
To: **Right atrium**

It arises as the fusion of the common iliac veins anterolateral to the L5 vertebral body lying posterior to the right common iliac artery. It ascends, initially posterolateral to, then lateral to the aorta and lies anterolateral to the right of the bodies of L5–L1. It lies on (from below up): right lumbar arteries, right renal artery, right sympathetic chain, right suprarenal gland, right crus of the diaphragm and right inferior phrenic artery. It is crossed by (from below up): the root of the ileal mesentery, the third part of the duodenum, the head of the pancreas, the bile duct, the portal vein, the first part of the duodenum, the posterior abdominal peritoneum and the bare area of the liver. It is hugged closely on its right side by the right suprarenal gland and forms the posterior wall of the epiploic foramen of Winslow below this. After passing through the caval orifice in the central tendon of the diaphragm (T8) with the right phrenic nerve lateral to it, it almost immediately enters the pericardium and then the inferior aspect of the right atrium. The vein possesses a 'valve-like' flap guarding the medial portion of its orifice.

Lumbar veins. Drain somewhat inconsistently but usually the 3rd and 4th drain directly into the inferior vena cava whilst above this level they drain into the ascending lumbar veins and hence to the azygos and hemiazygos systems. There are, however, usually connections of the 3rd and 4th lumbar veins with the ascending lumbar veins.

2

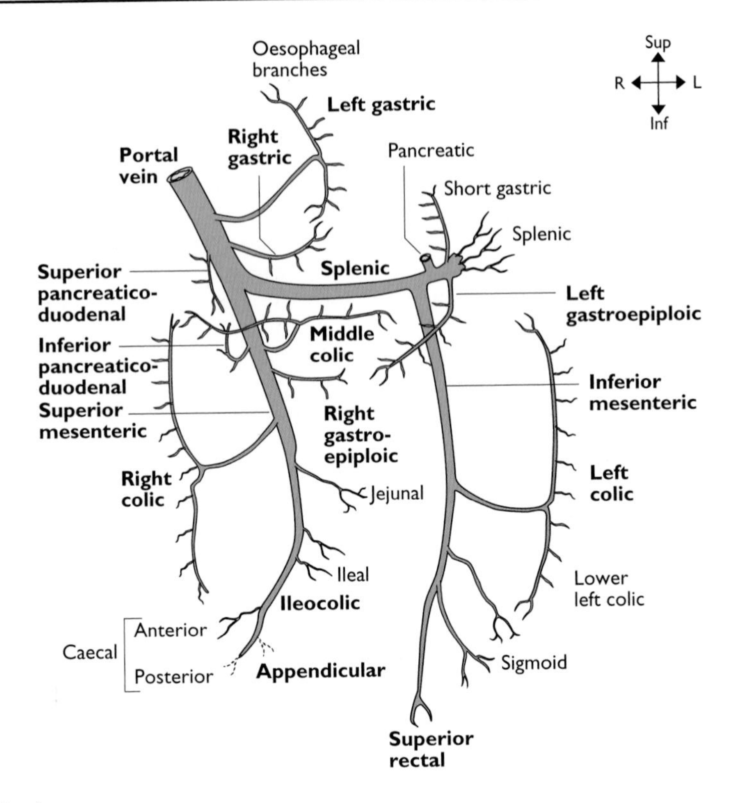

Portal vein

PORTAL VEIN
From: **Sup mesenteric & splenic Vs**
To: **Porta hepatis**

The vein is formed from the union of the superior mesenteric and splenic veins at the level of the disc between L1 and L2 just at the right of the midline. At its formation it lies posterior to the neck of the pancreas and anterior to the inferior vena cava. It runs superiorly inclining slightly to the right, lying posterior to the first part of the duodenum and anterior to the inferior vena cava. The bile duct comes to lie anterolateral to it from the right and the hepatic artery comes to lie anteromedial to it from the left. Ascending in the free border of the lesser omentum it continues to lie posterior to these two structures and forms the anterior margin of the aditus to the lesser sac (foramen of Winslow). It divides into terminal right and left branches as it enters the porta hepatis.

Portosystemic anastomoses

1 Lower end of oesophagus. The veins from the lower third of the oesophagus drain downwards to the left gastric vein (portal) and, above this level, oesophageal veins drain to the azygos and hemiazygos systems (systemic).

2 Upper end of anal canal. At the anal columns in the upper half of the anal canal there is a venous watershed between the drainage above by the superior rectal veins (portal via inferior mesenteric vein) and drainage below by the inferior and middle rectal veins (systemic via the pudendal and internal iliac veins).

3 Bare area of liver. Where the bare area of the liver lies in contact with the diaphragm there is a watershed of venous drainage between the hepatic veins (portal) and the inferior phrenic veins (systemic).

4 Peri-umbilical. The ligamentum teres represents a venous watershed in that its inner portion is drained into the portal system (via the left branch of the portal vein) whilst its outer portion is drained into the systemic system (indirectly via the great saphenous and axillary veins).

5 Retroperitoneal. Branches of the left and right colic and splenic veins (portal) may meet branches of the lumbar veins (systemic via inferior vena cava and azygos systems) in the retroperitoneal area.

3: LYMPHATICS

Note: Deep lymphatics follow arteries and superficial ones follow veins.

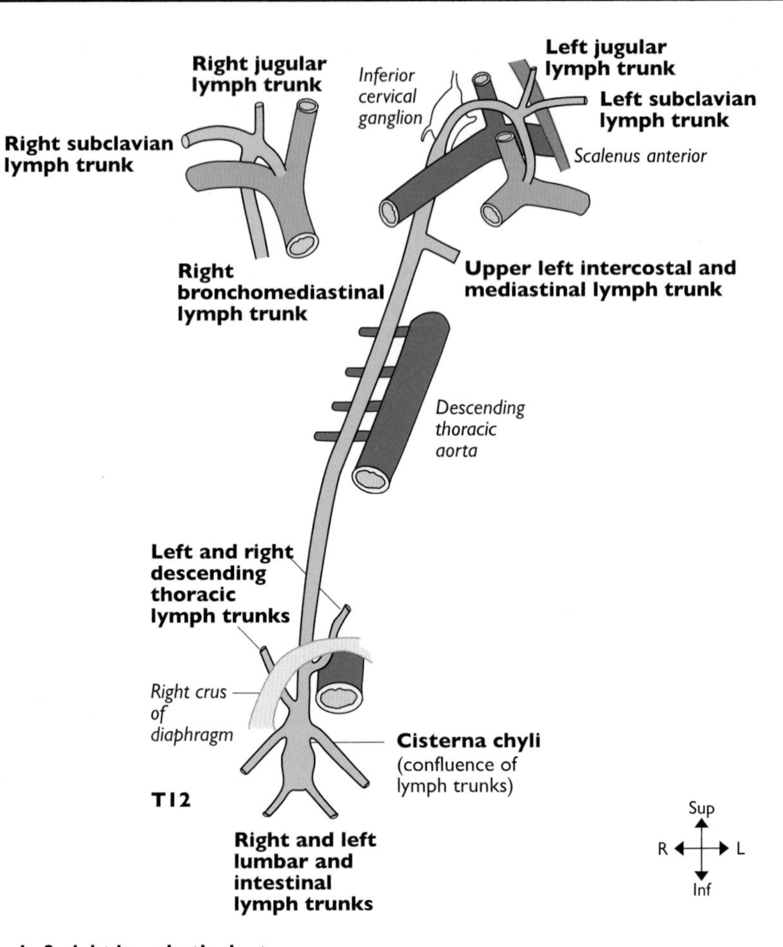

Thoracic & right lymphatic ducts

THORACIC & RIGHT LYMPHATIC DUCTS

From: **Cisterna chyli**
To: **Left subclavian V**

Thoracic duct

Receives:
- Left jugular trunk
- Left subclavian trunk
- Cisterna chyli
- Most thoracic lymphatics (except right upper thorax)

Drains:
- All body tissue below the diaphragm
- Left arm
- Left head and neck
- Left thorax
- Lower right thorax

It originates from the upper cisterna chyli on the right anterolateral side of the body of T12, lying lateral to the abdominal aorta. It passes posterior to the right crus of the diaphragm and ascends on the right posterior intercostal arteries with the aorta on its left and the azygos vein on its right. It slopes to the left in the mid thorax crossing the vertebral column posterior to the oesophagus at the level of T5. It continues superiorly to the left of the vertebral column, posterolateral to the oesophagus, posteromedial to the upper mediastinal pleura and posterior to the initial part of the left subclavian artery. It then passes anterior to the inferior cervical (stellate) ganglion before arching anteriorly over the left vertebral and left subclavian arteries and the dome of the pleura to lie along the medial edge of scalenus anterior before reaching the posterosuperior aspect of the left subclavian vein as the latter joins the left internal jugular vein.

Right lymphatic duct

Receives:
- Right subclavian trunk
- Right jugular trunk
- Right bronchomediastinal trunk (right upper thorax)

Drains:
- Right head and neck
- Right arm
- Right upper thorax

The three trunks usually drain separately but the first two often join to give a right lymphatic duct which ends in the right subclavian vein. In which case it has a very short course from its formation anterior to scalenus anterior, passing over the dome of the cervical pleura to reach the right subclavian vein as the latter joins with the right internal jugular vein.

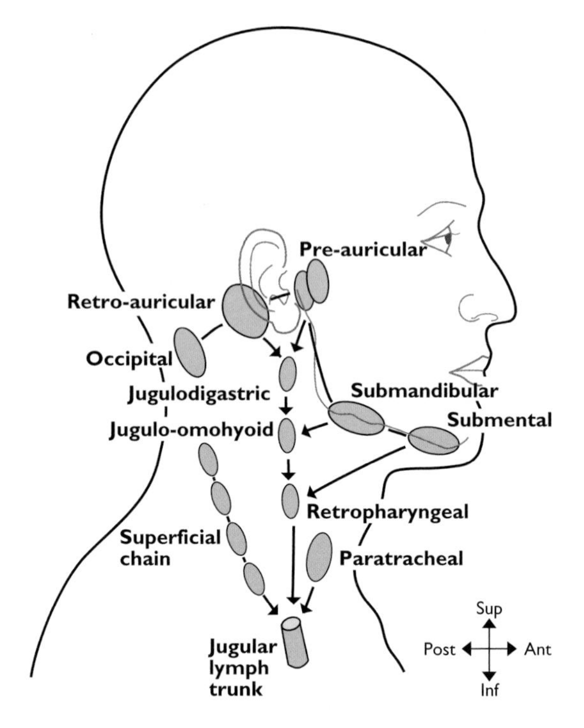

Lymph nodes — head & neck

LYMPH NODES—HEAD & NECK

Circular nodal chain

Submental (bilateral)
- Anterior tongue
- Floor of mouth
- Lower incisor and canine teeth
- Lower lip
- Skin of anterior chin

Submandibular
- Upper lip
- Cheek
- Nose
- Forehead and anterior scalp
- Middle tongue
- Lower molar and premolar teeth
- All upper teeth
- Sublingual gland
- Submandibular gland
- Anterior half of nasal cavity and nasal sinuses

Deep and superficial pre-auricular (parotid)
- Middle scalp
- Skin of temple
- Pinna
- Parotid gland
- Posterior orbit

Retro-auricular (mastoid)
- Pinna
- Posterior scalp

Occipital
- Posterior scalp

Deep cervical chain

Jugulodigastric
- Palatine tonsil
- Upper pharynx
- Posterior tongue

Jugulo-omohyoid
- Posterior half of nasal cavity and nasal sinuses
- Palate (hard and soft)

Retropharyngeal
- Pharynx

Paratracheal
- Hypopharynx
- Larynx
- Trachea
- Thyroid
- Parathyroids

Superficial cervical chain

Skin of neck

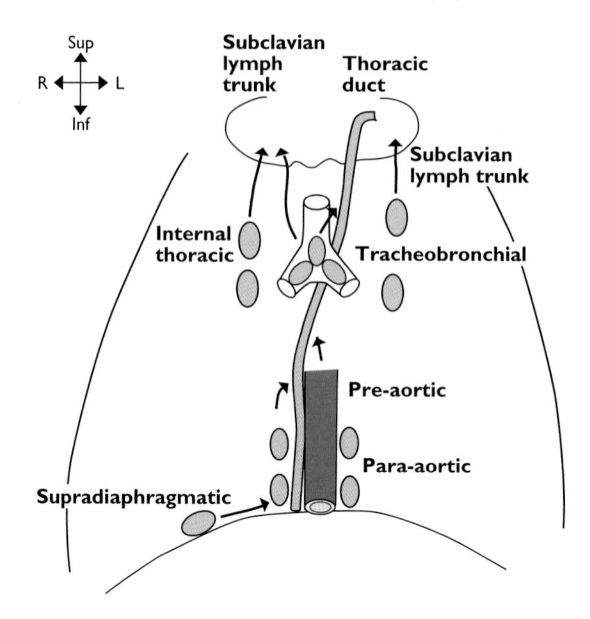

Lymph nodes—thorax

LYMPH NODES—THORAX

Pre-aortic
- Middle third of oesophagus

Supradiaphragmatic
- Diaphragm
- Subphrenic spaces
- Bare area of liver

Tracheobronchial
- Heart and all layers of pericardium
- Lungs and visceral pleura
- Extrapulmonary bronchi
- Trachea
- Thymus (occasionally thyroid isthmus)

Para-aortic
- Thoracic wall
- Parietal pleura
- Anterior abdominal wall

Internal thoracic
- Breast
- Anterior thoracic wall
- Upper abdominal muscles
- Diaphragm

Left and right lower thoracic nodes drain directly to the thoracic duct or via a separate left bronchiomediastinal trunk which joins the thoracic duct in the posterior superior mediastinum. Upper right thoracic nodes drain via the right bronchomediastinal trunk which either drains into the right lymphatic duct or directly into the right subclavian vein.

Note: Normal drainage from breast is to anterior and posterior axillary, infraclavicular and internal thoracic groups. With pathological blockage from disease the spread can be to opposite side, cervical, peritoneal cavity and liver, and inguinal nodes.

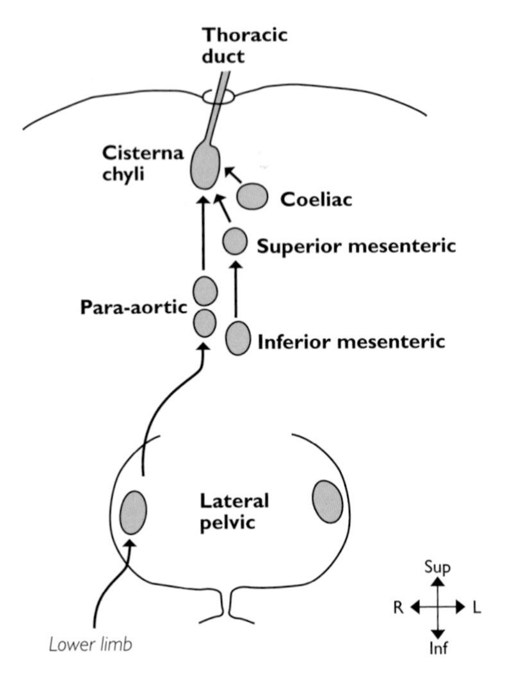

Lymph nodes—abdomen

LYMPH NODES—ABDOMEN

Coeliac
- Lower third of oesophagus
- Stomach and greater omentum
- First and upper second part of duodenum
- Spleen
- Pancreas
- Liver
- Gallbladder

Superior mesenteric
- Lower second, third and fourth parts of duodenum
- Jejunum
- Ileum
- Caecum and appendix
- Ascending colon
- Transverse colon

Inferior mesenteric
- Distal transverse colon
- Descending colon
- Sigmoid colon
- Upper rectum and rectal mucosa to dentate line

Para-aortic
- Inferior surface of diaphragm
- Suprarenal gland
- Kidney
- Gonad (plus fallopian tube in female)
- Superior lateral uterus
- Ureter
- Bare area of liver
- Posterior abdominal wall

Lateral pelvic nodes
- Lower rectum and dentate line
- Bladder
- Urethra
- Lower ureter
- Female
  - uterus
  - cervix
  - upper vagina
  - clitoris
  - labia minora
- Male
  - vas deferens
  - seminal vesicles
  - prostate
  - bulk of penis

Endodermal (gut-tube/gut-derived) structures drain to lymph nodes lying along their arteries of supply which are named according to the artery with which they are associated and are not listed individually. Their number and exact position are variable. However, the position and drainage of the highest node groups (listed above) are constant.

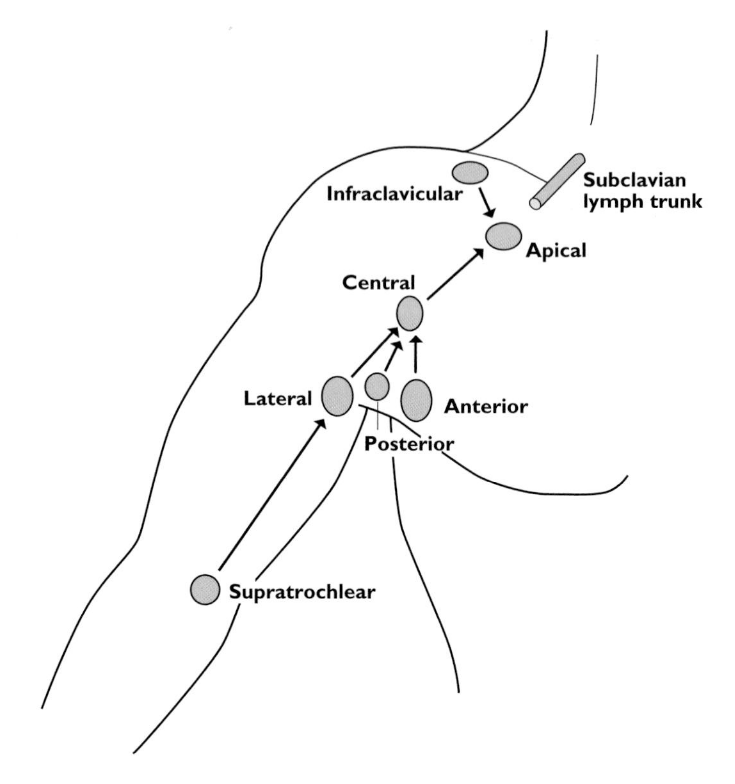

Lymph nodes—upper limb

LYMPH NODES—UPPER LIMB

Axillary groups

Anterior (pectoral)
- Breast
- Anterior thoracic wall
- Upper anterior abdominal wall

Posterior (subscapular)
- Posterior thoracic wall
- Tail of breast
- Upper posterior abdominal wall

Lateral
- Arm
- Forearm
- Hand

Central

Apical

Supratrochlear
- Skin of anterior forearm and hand

Infraclavicular
- Skin of shoulder
- Skin of lower neck
- Skin of anterior upper thoracic wall
- Breast

A useful mnemonic for the axillary nodes is the word APICAL—(A)nterior, (P)osterior, (I)nfraclavicular, (C)entral, (A)pical, (L)ateral

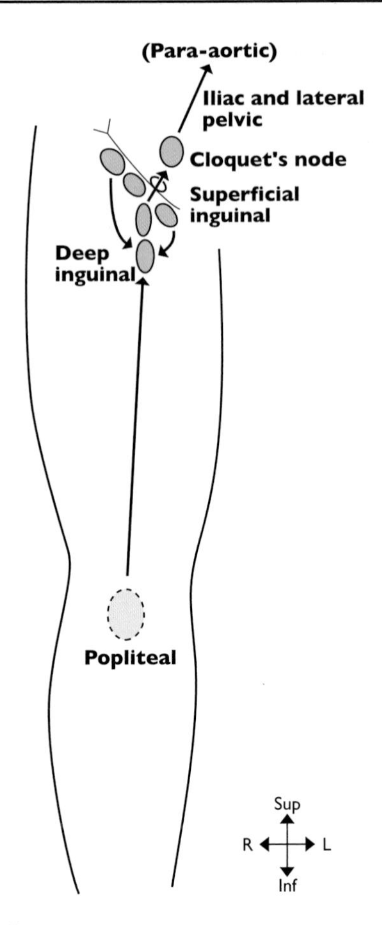

Lymph nodes—lower limb

LYMPH NODES—LOWER LIMB

Superficial inguinal
- Uterine fundus (via round ligament)
- Skin of penis
- Labia minora/scrotum
- Skin of buttock
- Skin of lower abdominal wall to umbilicus
- Skin of thigh, anterior skin of calf and dorsum of foot
- Skin of anterior perineum

Deep inguinal
- Anterior perineum
- Thigh
- Leg
- Foot

Popliteal
- Skin of sole of foot
- Skin of posterior calf

Note: Cloquet's node is the highest node of the lower limb and usually lies either inside the femoral canal or just at its superior limits.

Superficial inguinal lymphatics drain via the saphenous opening into the deep inguinal nodes and then via the femoral canal to Cloquet's node and onwards to the internal and external iliac nodes.

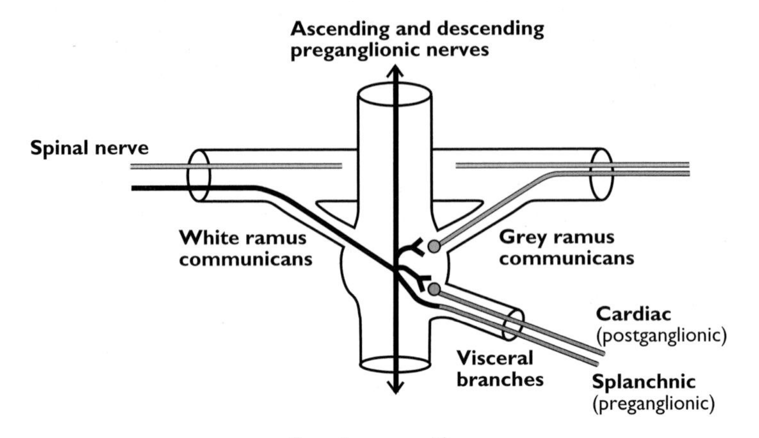

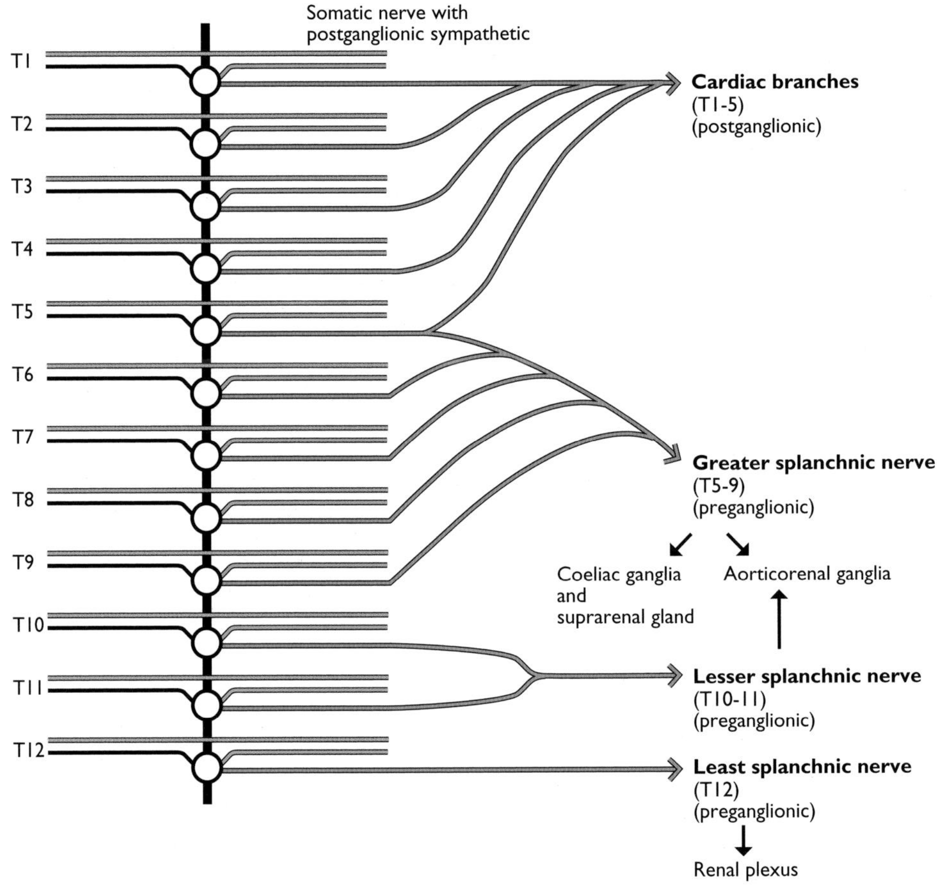

Thoracic sympathetics (T1–12)
Note: All splanchnic nerves synapse in collateral ganglia

4: AUTONOMIC NERVOUS SYSTEM

The autonomic nervous system, as the term implies, is an internal (visceral) adjustment system which is largely controlled automatically but is influenced, to some extent, by somatic activity and the special senses. It controls the activity of the viscera, glands, blood vessels, heart and smooth muscle.

The two parts, working complementarily, are the sympathetic and parasympathetic systems. The **sympathetic** system is active at times of urgent activity or stress, summarised as the fight, flight and fright response, at which time blood is needed rapidly in muscles and brain and can be diverted there from the skin and gut. There is, therefore, selective vasoconstriction, increase in heart rate and blood pressure together with bronchodilatation, decrease in intestinal peristalsis and closure of sphincters. In addition there are two further activities in the head and neck where the eyes open wide and the pupils dilate. On a less urgent basis somatic nerves to skin carry sympathetic fibres which constantly adjust sweating (sudomotor), blood flow (vasomotor) and hair erection (pilomotor) as a means of heat regulation.

Conversely, **parasympathetic** activity predominates during periods of quiet activity, rest and during and after feeding when the salivary glands secrete and food is digested. There is maximal intestinal activity and normal production of glandular secretions from mucosa throughout the body. The heart rate is slowed, blood pressure is minimised, the bronchi constricted to lessen the dead space of the airways and the pupils are constricted. An extra task for the parasympathetic is accommodation of the lens for near vision.

The efferent (motor) nerves of both systems are accompanied by general visceral afferent (sensory) fibres which detect visceral distension, excessive smooth muscle contraction, hunger, nausea and sexual excitement. From the heart and lungs sensation is probably in both sympathetic and parasympathetic systems whereas in the abdomen it is predominantly in the sympathetic system. Sensation from pelvic organs of cloacal origin travel with the parasympathetics.

Sympathetic

The following 10 'rules' summarise the general plan for the sympathetics:

1 Sympathetic motor function stimulates: sweating, pilo-erection and vascular constriction in the skin. Throughout the rest of the body it controls vascular calibre and smooth muscle tone, stimulating closure of sphincters (pylorus, internal anal, vesical). It decreases glandular secretion; causes ejaculation, bronchodilatation; and also increases the heart rate.

2 In addition to these it also has two special functions in the head and neck for dilating the pupil and raising the eyelids.

3 Sympathetic outflow from the spinal cord occurs only from preganglionic cell bodies in the lateral horns of T1 to L2. These myelinated preganglionic fibres (shown in black) exit with the somatic motor root from the ventral horn but soon leave the ventral ramus to pass to the ganglia of the sympathetic chain in a white ramus communicans.

4 Although the spinal connections to the sympathetic chain of ganglia are only between T1 and L2, the chain itself extends the whole length of the body with a ganglion at the level of each spinal nerve in the lumbosacral region and three ganglia above T1—superior, middle and inferior cervical ganglia. The inferior cervical ganglion is usually fused with the T1 ganglion and the two together are then termed the stellate ganglion.

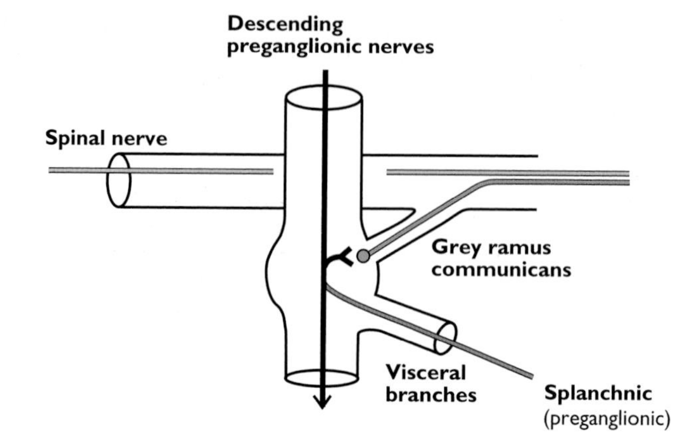

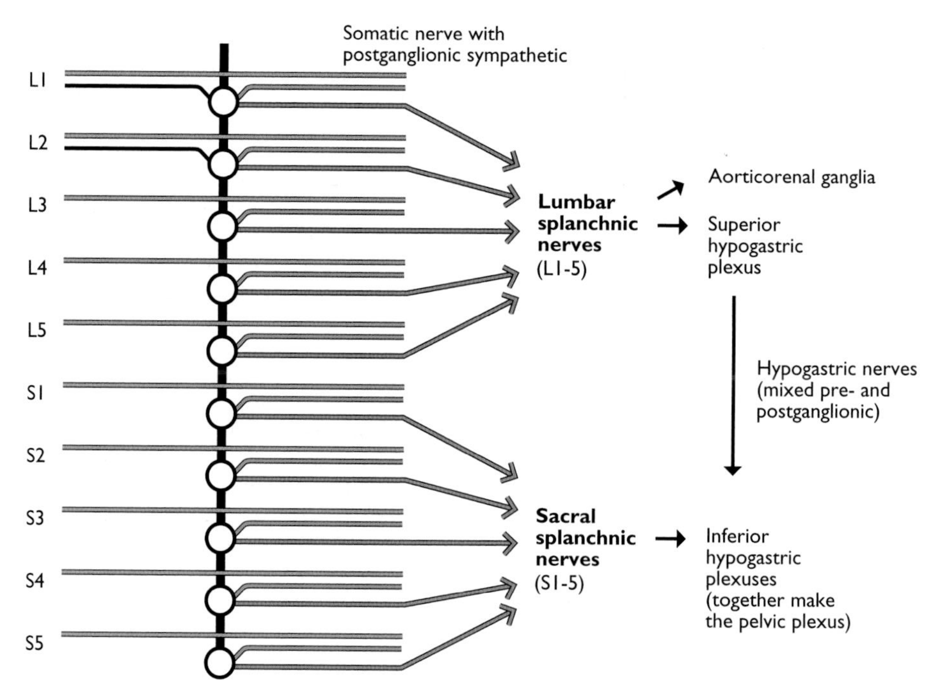

Lumbosacral sympathetics (L1–S5)

Note: L1–2 have white rami communicans. Lumbar and sacral splanchnic nerves are all preganglionic. They synapse in the inferior hypogastric plexus and perhaps some in the superior hypogastric plexus

5 When the preganglionic fibres reach a ganglion of the sympathetic chain they have two alternatives. They can leave that ganglion for distribution, with or without synapsing (see below). Alternatively, they can pass up or down the chain to a higher or lower ganglion before leaving for distribution, with or without synapsing (see below). Thus the upper thoracic sympathetics supply the arm and head and neck regions whilst the lower thoracic and upper lumbar sympathetics supply the lower trunk and leg.

6 In the ganglion from which the sympathetic is destined to leave, the preganglionic fibres always synapse and leave as postganglionic fibres *unless* they are supplying *gut* or *suprarenal gland* in which case they continue as preganglionic fibres.

7 Each ganglion has a somatic (lateral) branch from which unmyelinated postganglionic fibres (shown in green) pass into a grey ramus communicans to reach the somatic nerve at the same level.

8 Each ganglion has a visceral (medial) branch (shown in red) which supplies internal organs, such as the heart, with unmyelinated postganglionic fibres. Note that fibres to the gut and suprarenal gland, in the form of greater, lesser, least, lumbar and sacral splanchnic nerves, are all preganglionic (see 6 above). They synapse in collateral ganglia nearer the organ of distribution (coeliac, renal, inferior hypogastric, etc.) before becoming postganglionic. The medulla of the suprarenal gland can be considered as a ganglion.

9 In addition to the somatic and visceral branches, the three cervical ganglia have vascular (anterior) branches (shown in grey) which are distributed as in the diagram. These vascular branches simply allow a wider distribution of sympathetics to major vessels and into the skull than is possible via somatic or visceral branches. For the special function of pupillary dilatation and eyelid raising these sympathetics are distributed via cranial nerves—ophthalmic division of trigeminal and oculomotor respectively.

10 Sympathetic afferent (sensory) fibres (not shown) from the viscera return along the visceral efferent branches, pass through the ganglia of the sympathetic chain and the white rami communicantes before reaching their cell bodies in the dorsal root ganglia. They enter the spinal cord with the somatic sensory fibres via the dorsal horn. Note that the grey rami communicantes contain postganglionic efferent sympathetics only whilst the white rami communicates contain both preganglionic efferent sympathetic fibres and afferent sympathetic fibres.

4

Parasympathetic

The following 10 'rules' summarize the general plan for the parasympathetics:

1 Parasympathetic motor function stimulates: glandular secretions, particularly salivary glands and mucous glands throughout the body; peristalsis; opening of sphincters (pylorus and internal anal); penile erection; slowing of the heart; and bronchial constriction.

2 Two special functions in the head and neck are pupillary constriction and accommodation of the lens for near vision.

3 The only parasympathetic outflow in the body is with the cranial nerves III, VII, IX, X and from sacral segments S2,3,4.

4 All myelinated preganglionic efferent fibres (shown in red) that are destined to supply structures in the head and neck emerge from the brain stem with fibres of the relevant cranial nerve (III, VII and IX) and run to specific parasympathetic ganglia where they synapse (see diagram, see p. 78).

5 Each of these peripheral parasympathetic ganglia (ciliary, pterygopalatine, submandibular and otic) have a characteristic pattern of nerve connection in that in addition to the parasympathetic synapse, there is a sensory branch of the trigeminal nerve (V) and vasomotor fibres of the sympathetic both passing straight through the ganglion (see diagram, see p. 79). Beyond the ganglion, the sensory branch of the trigeminal nerve carries the postganglionic, unmyelinated parasympathetic fibres (also shown in red) and the sympathetic fibres to the end organ.

6 Each vagus nerve (X) has two cervical parasympathetic branches that pass down to the heart and then each vagus continues to

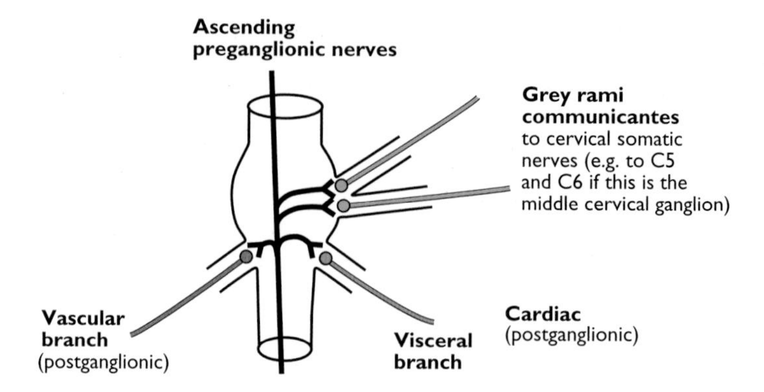

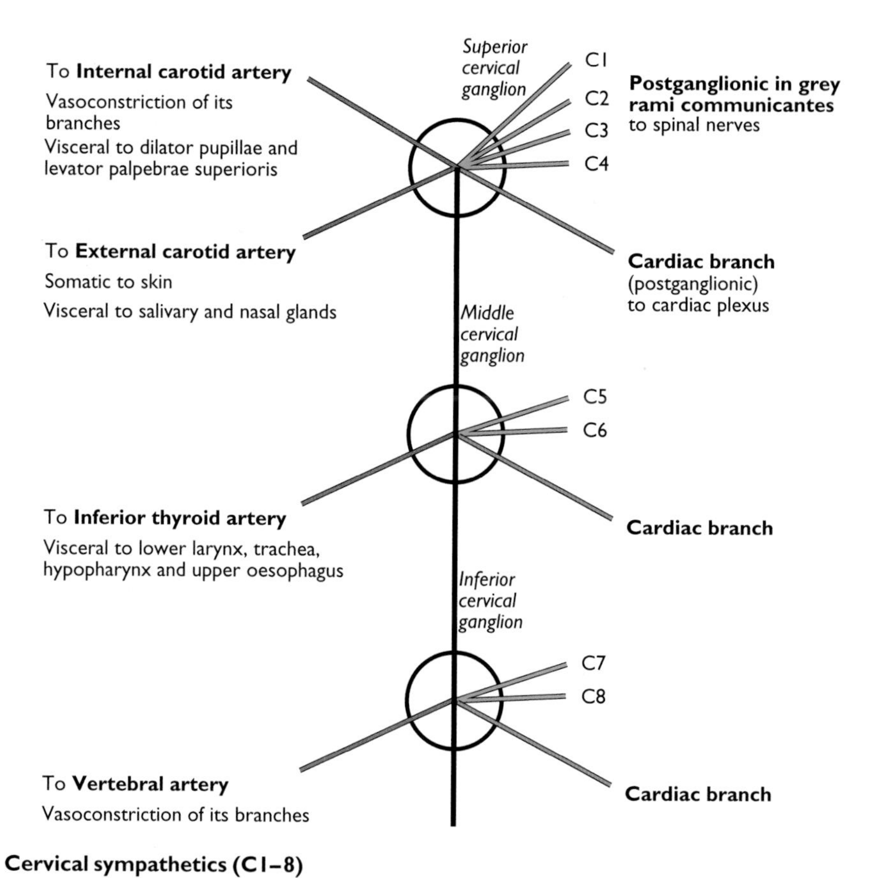

Cervical sympathetics (C1–8)

supply parasympathetic fibres to thoracic and abdominal structures as far as two-thirds along the transverse colon. The preganglionic fibres synapse in peripheral ganglia so that the postganglionic fibres are usually short.

7 The parasympathetic outflow in the sacral region (S2,3,4) arises in the lateral grey horn but emerges via the ventral horn and the fibres constitute a pelvic splanchnic nerve. Each nerve passes to its inferior hypogastric plexus (left and right together making the pelvic plexus) and then to the walls of the pelvic viscera where these preganglionic fibres synapse in small ganglia before distribution. On the left, postganglionic fibres pass superiorly via the hypogastric nerves to the superior hypogastric plexuses thence to the left third of the transverse, left and sigmoid colons, and the rectum via the branches of the inferior mesenteric artery; some fibres run in the tissue of the left retroperitoneum and mesentery.

8 Although not strictly part of the parasympathetic system, there are afferent (sensory) fibres running with the parasympathetics. In the vagus (X) there are general visceral afferent fibres from the thorax and abdomen (transmitting pain, visceral distension, etc.) which have their cell bodies in the inferior vagal ganglion and their central connections probably in the nucleus solitarius or dorsal vagal nucleus. For pelvic sensation similar fibres reach sacral segments S2,3,4, with their cell bodies in the dorsal root ganglia before they enter the dorsal horn of the spinal cord.

9 Other sensory fibres running with the parasympathetics are special visceral afferents which detect taste and changes in the baro- and chemoreceptors in the carotid sinus and body respectively.

10 The carotid sinus and body are supplied by both the glossopharyngeal nerve (IX) and the vagus (X) with cell bodies in their respective inferior ganglia. Taste is carried in VII, IX and X. The cell bodies for taste in VII are in the geniculate ganglion, those in IX are the inferior glossopharyngeal ganglion, whilst those in X are in the inferior vagal ganglion. Central connections are in the nucleus solitarius (see summary table of cranial nerve nuclei and fibres).

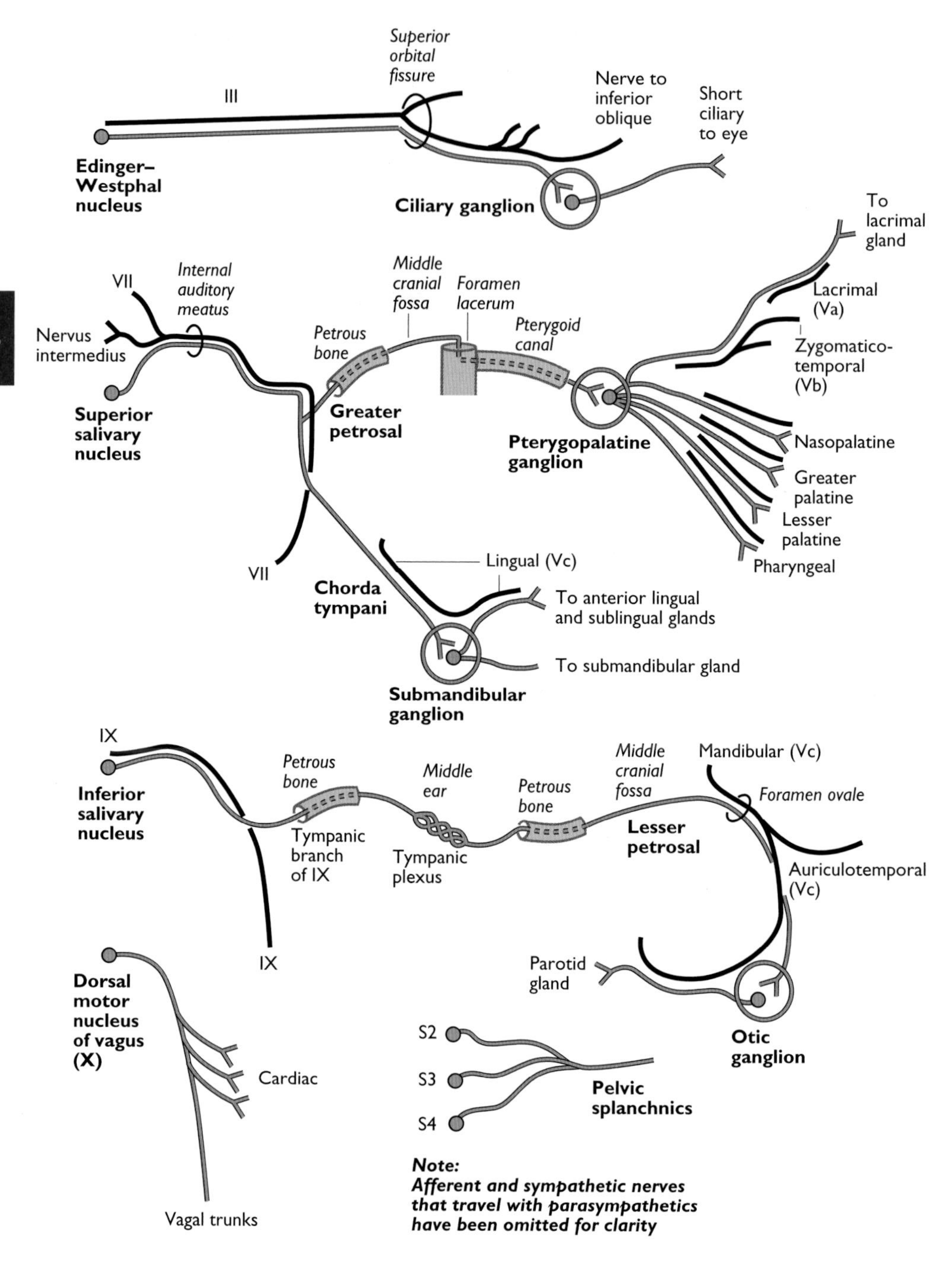

Note:
Afferent and sympathetic nerves that travel with parasympathetics have been omitted for clarity

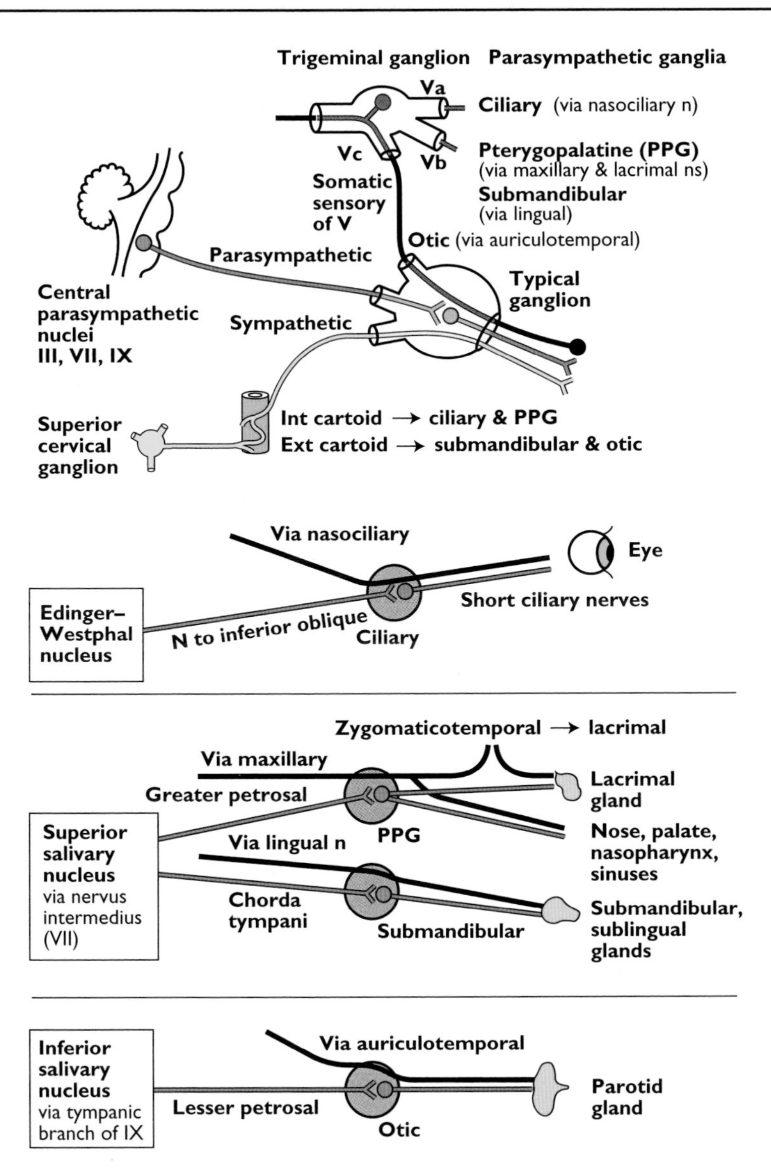

Typical and specific connections of the parasympathetic ganglia
PPG: pterygopalatine ganglion

Pathways for cranial outflow of parasympathetic nerves

Cranial nerve	Central nucleus	Nerve carrying preganglionic fibres	Pathway & foramen	Site of ganglion	Name of ganglion	Nerve carrying postganglionic fibres	Organs supplied
III	**Edinger–Westphal** (midbrain)	III→ Nerve to inferior oblique	Cavernous sinus→ superior orbital fissure→ orbit	Between optic N & lat rectus in apex of orbit	**Ciliary**	**Va** Nasociliary & short ciliary	**Ciliary muscle for accommodation. Circular muscle of pupil for constriction**
VII	**Superior salivary** (pons)	Nervus intermedius → VII→ greater petrosal N→ N of pterygoid canal	Internal acoustic meatus→ middle ear→ middle cranial fossa → pterygoid canal	Pterygopalatine fossa	**Pterygopalatine**	**Vb** (maxillary branches)	**Mucosal glands of nose, nasopharynx, sinuses, soft palate**
						Vb (infra-orbital → zygomatic → z-temporal) → **Va** (lacrimal)	**Lacrimal gland**
VII	**Superior salivary** (pons)	Nervus intermedius → VII→ chorda tympani → lingual N	Internal acoustic meatus→ middle ear → petrotympanic fissure→ infratemporal fossa	Below lingual N on hyoglossus	**Submandibular**	**Vc** (lingual)	**Submandibular, sublingual, ant. lingual, mucosal glands on side of tongue**
IX	**Inferior salivary** (medulla)	IX→ tympanic branch → lesser petrosal N	Middle ear →middle cranial fossa → f. ovale	Below f. ovale on N to tensor tympani & palati	**Otic**	**Vc** (auriculotemporal)	**Parotid gland**
IX	**Inferior salivary** (medulla)	Pharyngeal & laryngeal branches	Direct to oropharynx & post third of tongue	In mucosa of oropharynx & post third of tongue			**Mucosal glands of oropharynx, post third of tongue**
X	**Dorsal motor** (medulla)	X (Vagus)	Cardiac branches from neck→ plexus on oesophagus → abdomen	On target organs			**All viscera of thorax & abdomen → 2/3 across transverse colon (S2,3,4 for rest)**

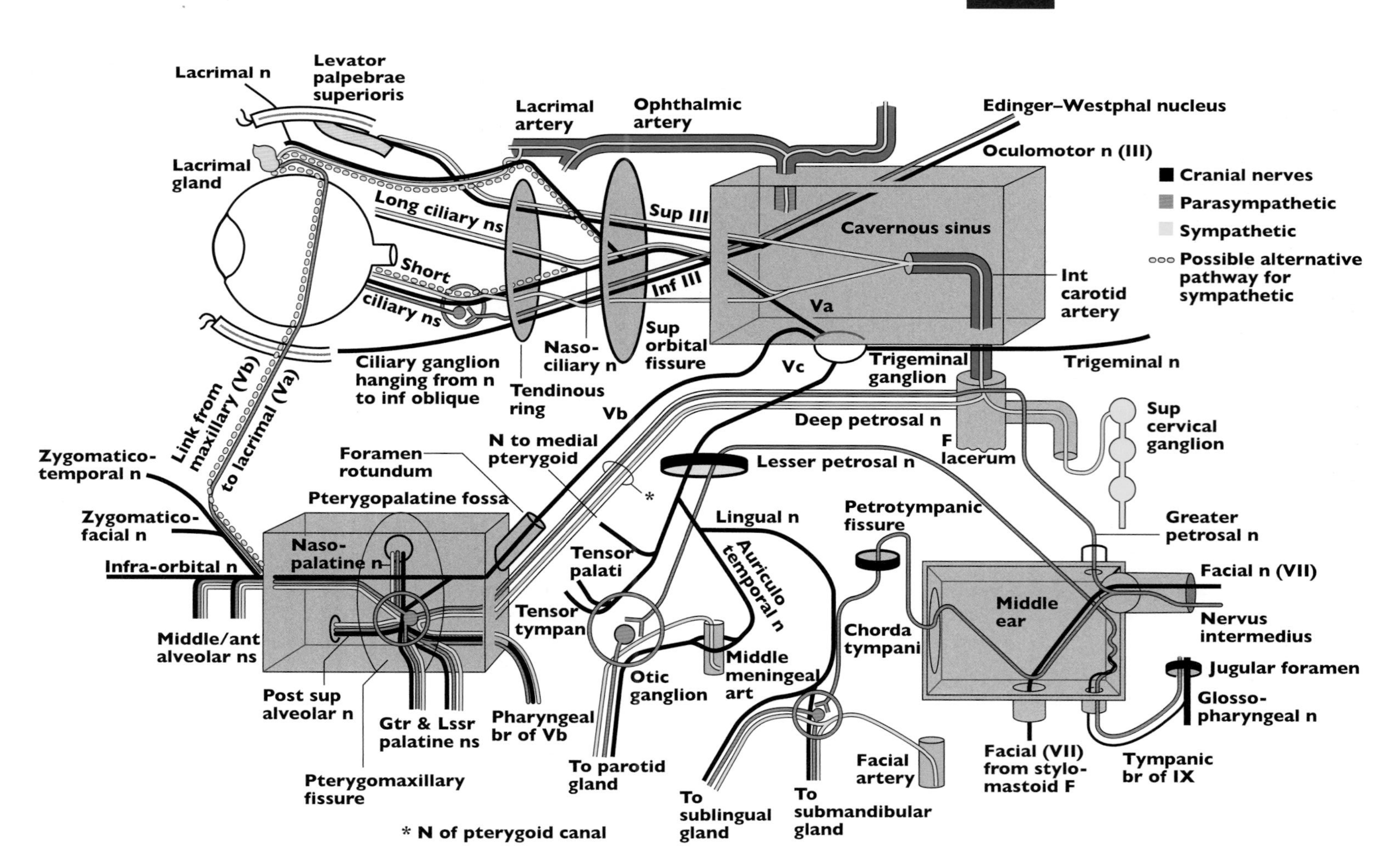
Lacrimal n
Levator palpebrae superioris
Lacrimal artery
Ophthalmic artery
Edinger–Westphal nucleus
Oculomotor n (III)
Cranial nerves
Parasympathetic
Sympathetic
Possible alternative pathway for sympathetic
Lacrimal gland
Long ciliary ns
Sup III
Cavernous sinus
Short ciliary ns
Inf III
Int carotid artery
Va
Sup orbital fissure
Naso-ciliary n
Ciliary ganglion hanging from n to inf oblique
Tendinous ring
Vc
Trigeminal ganglion
Trigeminal n
Link from maxillary (Vb) to lacrimal (Va)
Vb
Deep petrosal n
F lacerum
Sup cervical ganglion
Zygomatico-temporal n
Foramen rotundum
N to medial pterygoid
Lesser petrosal n
*
Pterygopalatine fossa
Zygomatico-facial n
Lingual n
Petrotympanic fissure
Greater petrosal n
Naso-palatine n
Infra-orbital n
Tensor palati
Auriculo temporal n
Facial n (VII)
Middle ear
Tensor tympan
Chorda tympani
Nervus intermedius
Middle/ant alveolar ns
Middle meningeal art
Otic ganglion
Jugular foramen
Post sup alveolar n
Glosso-pharyngeal n
Gtr & Lssr palatine ns
Pharyngeal br of Vb
Facial (VII) from stylo-mastoid F
Tympanic br of IX
To parotid gland
Facial artery
Pterygomaxillary fissure
To sublingual gland
To submandibular gland
* N of pterygoid canal

Summary of taste

The sensation of taste originates in taste buds in the mucosa of the tongue and oropharynx. The buds are surrounded by the endings of the gustatory nerves which transmit taste as special visceral afferent sensation to the nucleus of the tractus solitarius in the brain stem. The three cranial nerves involved are facial (VII), glossopharyngeal (IX) and vagus (X) and the cell bodies for taste are in the geniculate, inferior glossopharyngeal and inferior vagal ganglia respectively.

Taste buds are found as follows:

1 As single buds in the mucosa.

2 In fungiform papillae on the anterior two-thirds of the tongue.

3 In the grooves around vallate papillae which are just anterior to the sulcus terminalis of the tongue.

4 In foliate papillae on the sides of the anterior two-thirds of the tongue.

Position of taste bud	Type of bud (1, 2, 3, 4 as above)	Nerve supply
Anterior two-thirds of tongue	1, 2, 4	Chorda tympani (VII) Carried by lingual N
Associated with posterior third of tongue	1, 3	Glossopharyngeal
Posterior wall of oropharynx	1	Glossopharyngeal
Palatoglossal arches	1	Glossopharyngeal
Oral aspect of soft palate	1	Lesser palatine (maxillary division of trigeminal) and glossopharyngeal
Anterior surface of epiglottis and valleculae	1	Internal laryngeal branch of superior laryngeal nerve (vagus)

Distribution of autonomics of head and neck

This complex for "reference only diagram" is for those who want to see the global arrangement of the autonomics in the neck and head. Parasympathetics are shown in red. Cranial nerves carrying the pre- or postganglionic fibres are shown in black. Sympathetics are shown in yellow. Interrupted yellow lines indicate probable additional pathways for sympathetics (*see p. 82, left hand page*)

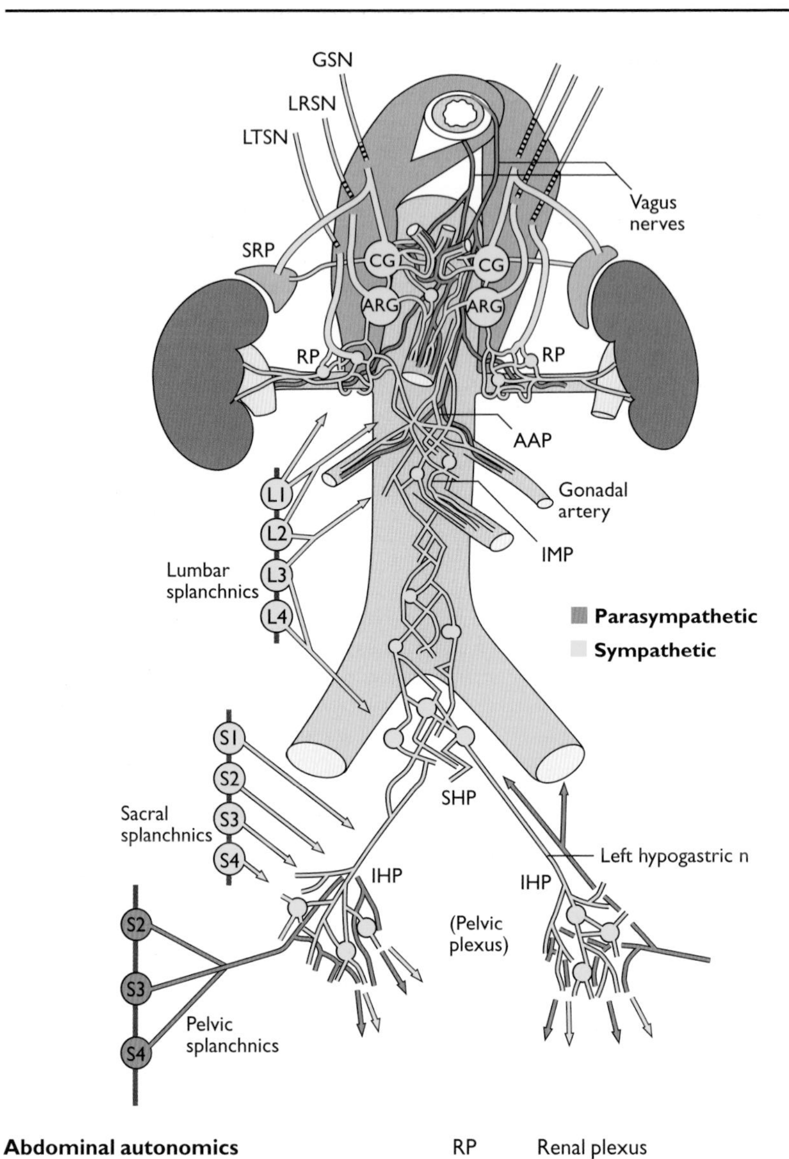

Abdominal autonomics

GSN	Greater splanchnic nerve
LRSN	Lesser splanchnic nerve
LTSN	Least splanchnic nerve
CG	Coeliac ganglion
ARG	Aorticorenal ganglion
RP	Renal plexus
SRP	Suprarenal plexus
AAP	Abdominal aortic plexus
IMP	Inferior mesenteric plexus
SHP	Superior hypogastric plexus
IHP	Inferior hypogastric plexus

Abdominal autonomics

The **greater splanchnic nerve** (T5–9) supplies the coeliac and aorticorenal ganglia and the suprarenal gland with preganglionic sympathetic fibres.

The **lesser splanchnic nerve** (T10,11) supplies the aorticorenal ganglion with similar fibres.

The **least splanchnic nerve** (T12) supplies the renal plexus with similar fibres.

Each of the splanchnic nerves pierces the crura of the diaphragm to enter the abdomen. They each carry efferent and afferent fibres.

The **coeliac ganglia** lie on each side of the coeliac trunk. They are supplied by preganglionic sympathetic fibres from the greater splanchnic nerves. Postganglionic sympathetic fibres leave these ganglia and form the coeliac plexus.

The **coeliac plexus** connects the coeliac ganglia across the midline; it surrounds the coeliac truck and extends down to become the superior mesenteric plexus. The coeliac plexus also receives preganglionic parasympathetic fibres from the vagus nerves. The fate of these fibres, both sympathetic and parasympathetic, is twofold. Many leave the plexus on branches of the coeliac trunk to be distributed to the bowel and other organs such as liver and spleen. Others pass downwards to reach other plexuses before being distributed similarly.

The **aorticorenal ganglia** are partially detached parts of the coeliac ganglia, lying just inferiorly. They contribute to both the coeliac and renal plexuses. As with the coeliac ganglia, sympathetic fibres leaving these ganglia are postganglionic.

A **renal plexus** surrounds the beginning of each renal artery. It has postganglionic sympathetic contributions from the coeliac and aorticorenal ganglia and preganglionic contributions from the least splanchnic nerve and the first lumbar splanchnic nerve. It has a few small ganglia for these preganglionic fibres to synapse. In addition it has both sympathetic and parasympathetic contributions from the coeliac and abdominal aortic plexuses. It supplies mainly the kidney and upper ureter with sympathetic and parasympathetic fibres although the function of the latter fibres is not clear.

The **superior mesenteric plexus** around the superior mesenteric artery is a downwards extension of the coeliac plexus. Its mixed sympathetic and parasympathetic fibres are distributed on this artery.

The **abdominal aortic (intermesenteric) plexus** lies on the aorta between the superior and inferior mesenteric arteries. It is connected above to the coeliac ganglia and plexus, aorticorenal ganglia and vagus nerves; below, it is in continuity with the inferior mesenteric and superior hypogastric plexuses. It also has preganglionic sympathetic input from the first and second lumbar splanchnic nerves.

The **vagus nerves** enter the abdomen via the oesophageal opening and distribute to abdominal organs and the bowel as far as two-thirds along the transverse colon via the coeliac and superior mesenteric plexuses. These preganglionic fibres synapse in small ganglia in the walls of the organs or bowel.

The **inferior mesenteric plexus** surrounds the beginning of the inferior mesenteric artery and is supplied by the abdominal aortic plexus with additional preganglionic sympathetic input from the second and third lumbar splanchnic nerves. Parasympathetic fibres from the sacral outflow (S2,3,4) ascend via the left inferior and superior hypogastric plexuses to be distributed with the sympathetic fibres on branches of the inferior mesenteric artery.

The **suprarenal plexus** on each side is supplied by the greater splanchnic nerve and a branch from the coeliac ganglion. The majority of sympathetic fibres reaching it are preganglionic to the medulla. There is no parasympathetic supply to the suprarenal gland.

The **lumbar splanchnics** are sympathetic preganglionic fibres that leave the sympathetic chain to supplement the thoracic splanchnics. L1 joins the renal plexus, L1, L2 the abdominal aortic plexus, L2,L3 the inferior mesenteric plexus and L3,L4 the superior hypogastric plexus.

The **superior hypogastric plexus** lies over and just below the bifurcation of the aorta. It is supplied by fibres continuing down from the abdominal aortic plexus (postganglionic) and the third and fourth lumbar splanchnic nerves (preganglionic). It contains a few small ganglia before forming **left and right hypogastric nerves** which pass down to the inferior hypogastric plexuses. It supplies the iliac vessels via the iliac plexuses and the ureter. It also has pelvic parasympathetics (S2,3,4) ascending through it on the way to the inferior mesenteric artery to supply bowel from the left side of the transverse colon downwards.

There is an **inferior hypogastric plexus** on each side wall of the pelvis. The two together make the pelvic plexus. They are supplied by pre- and postganglionic sympathetic fibres running in the hypogastric nerves from the superior hypogastric plexus and are supplemented by the first and second sacral splanchnics (preganglionic sympathetic) and the pelvic splanchnics (parasympathetic). They contain small ganglia for the synapses of any remaining preganglionic sympathetic fibres. The postganglionic sympathetic outflow from this plexus runs on arteries to give vasomotor supply and motor fibres to vas, seminal vesicles, prostate, anal and bladder sphincters and to inhibit peristalsis. Sympathetic sensory fibres arise in the upper rectum and body of uterus.

The **sacral splanchnics** are sympathetic preganglionic fibres that leave the sympathetic chain to supplement the pelvic sympathetics. S1 and S2 join the pelvic plexus or hypogastric nerve on each side. S3 and S4 from each side form a plexus on the median sacral artery.

The **pelvic splanchnics** are the caudal parasympathetic outflow (S2,3,4) and these fibres join the inferior hypogastric plexuses before distribution either on arteries or lying free in the retroperitoneum. The fibres are preganglionic and synapse in the walls of the organs they supply. They are motor to large bowel beyond the left third of the transverse colon, bladder and uterus. They are also sensory to some pelvic organs (bladder, proximal urethra, ampulla of rectum, anal canal, cervix of uterus, upper vagina and prostate). Ascending mostly from the left inferior hypogastric plexus, are those fibres mentioned above that supply parasympathetics to the left large bowel beyond the distribution of the vagus. Some of these fibres pass through the superior hypogastric plexus and then the inferior mesenteric plexus to reach the inferior mesenteric artery, whilst others may run directly to the left colon via the retroperitoneum.

Notes:

1 The neurotransmitter substance is acetylcholine at the endings of all preganglionic autonomic nerves whether sympathetic or parasympathetic and also at the postganglionic parasympathetic endings. All these endings are termed cholinergic. All postganglionic sympathetic endings have either noradrenalin or adrenalin as the neurotransmitter except sweat glands which are cholinergic.

2 Unless mentioned above as specific actions, the sympathetic efferent (motor) fibres cause vasoconstriction, inhibit intestinal peristalsis and glandular activity. The parasympathetic efferent (motor) fibres, however, cause glandular secretion and intestinal peristalsis but are inhibitory to the pyloric and ileocaecal sphincters. There are also specific actions of penile/clitoral erection and contraction of the bladder and uterus.

5: CRANIAL NERVES

Note: Apart from cranial nerves I and II, which clearly arise from specific sensory areas and their fibres pass directly to the brain, a nomenclature has been used to describe each nerve as arising centrally in the brain and passing out to its terminal branches irrespective of whether the nerve carries sensory, motor or a combination of fibres. Sympathetics have generally been omitted except in association with cranial nerves III and Va where they have specific functions and are carried by a branch of the cranial nerve.

Colour coding in figures

- Somatic motor—black
- Somatic sensory—black
- Special visceral motor (branchial muscles) —blue
- Special visceral sensory (taste and arterial receptors)—green
- General visceral motor (parasympathetic)—red
- General visceral sensory (parasympathetic)—green
- Special senses—black
- Sympathetic—yellow

Summary table of nuclei & fibres of cranial nerves

	Somatic motor	Special visceral motor (branchial)	General visceral motor (parasympathetic)	General visceral sensory	Special visceral sensory	Somatic sensory	Special senses
I							Smell **Limbic system**
II							Sight **Lat geniculate body**
III	**Nu: Oculomotor** Recti (sup, med, inf), inf oblique, levator palpebrae superioris		**Nu: Edinger–Westphal** *Ciliary ganglion* Ciliary body & muscle, sphincter pupillae				
IV	**Nu: Trochlear** Sup oblique						
V		**Nu: Motor of trigeminal** M of mastication, mylohyoid, ant digastric, tensors palati & tympani				**Nu: Sensory of V** Mesencephalic: proprioception Main: touch Spinal: pain & temperature For V (face, orbit, tongue)	
VI	**Nu: Abducent** Lat rectus						
VII		**Nu: Facial** M of facial expression, buccinator, post digastric, stylohyoid, stapedius	**Nu: Sup salivary** *Pterygopalatine & submandibular ganglia* Lacrimal, submandibular, sublingual & palatine glands		**Nu: Solitarius** Chorda tympani Taste: ant 2/3 tongue	**Nu: Sensory of V** Some skin of ext auditory meatus & tympanic membrane	
VIII							Hearing: **2 Nu** Equilib: **4 Nu**
IX		**Nu: Ambiguus** Stylopharyngeus	**Nu: Inf salivary** *Otic ganglion* Parotid, glands in post 1/3 tongue & oropharynx		**Nu: Solitarius** Taste: post 1/3 tongue, vallate papillae, oropharynx; baro- & chemoreceptors	**Nu: Sensory of V** Post 1/3 tongue, palate, pharynx, tonsil, middle ear	
X		**Nu: Ambiguus** M of pharynx, upper oesophagus, palate, larynx (from cranial XI)	**Nu: Dorsal motor of vagus** Cardiac & visceral muscle in thorax & abdomen	**Nu: Solitarius or dorsal sensory of vagus** From heart, lungs & abdominal viscera	**Nu: Solitarius** Taste: vallecula & epiglottis; baro- & chemoreceptors	**Nu: Sensory of V** Small area of skin behind auricle, in ext auditory meatus & on ear drum; pharynx; larynx	
XI	**Nu: Lat roots C1–5** Sternocleidomastoid & trapezius	**Nu: Ambiguus** M of palate & pharynx via vagus					
XII	**Nu: Hypoglossal** M of tongue (not palatoglossus)						
NB	Motor nuclei for skeletal muscle Cell bodies in CNS	Motor to muscles of branchial origin. Cell bodies in CNS	Motor nerves synapse in parasympathetic ganglia	From heart, lungs & gut	Taste; baro- & chemoreceptors	From skin & membranes. Cell bodies outside CNS except mesencephalic nu.	
				All sensory fibres relay to: motor nuclei, cerebellum, opposite thalamus → sensory cortex			

CNS = central nervous system; M = muscles; Nu = nucleus

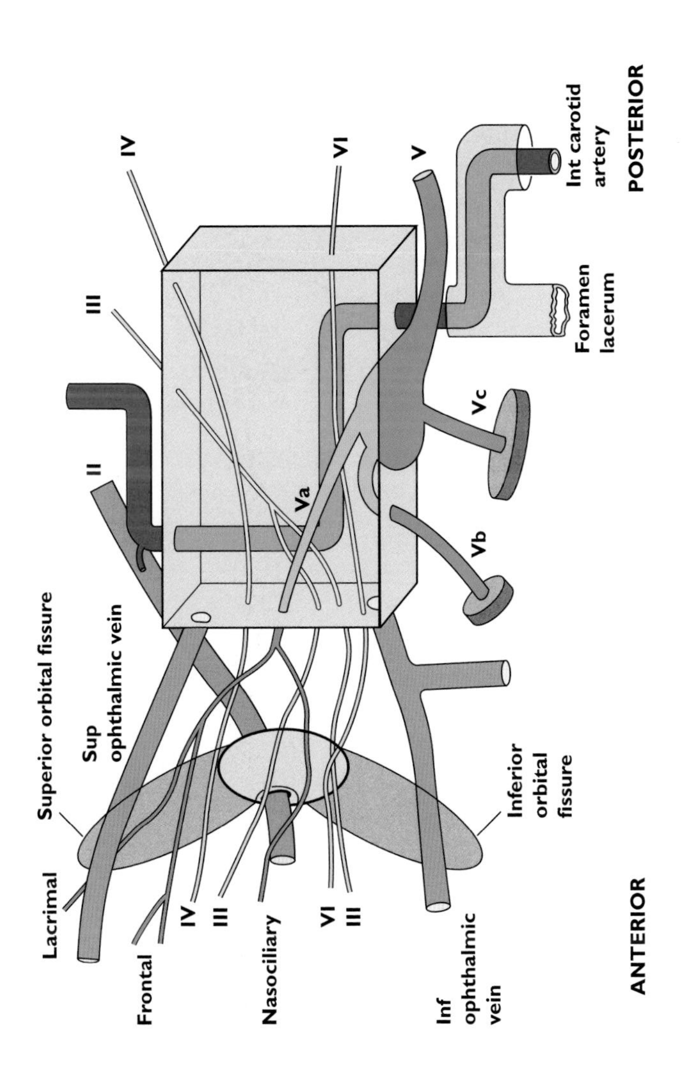

Lateral view of left cavernous sinus as a simplified box shape

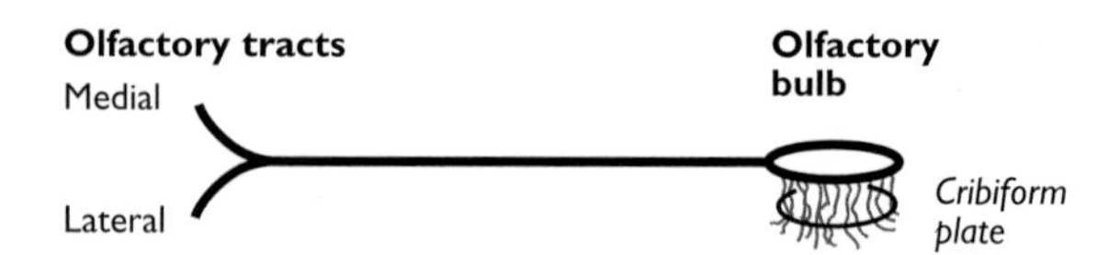

Olfactory nerve (I)

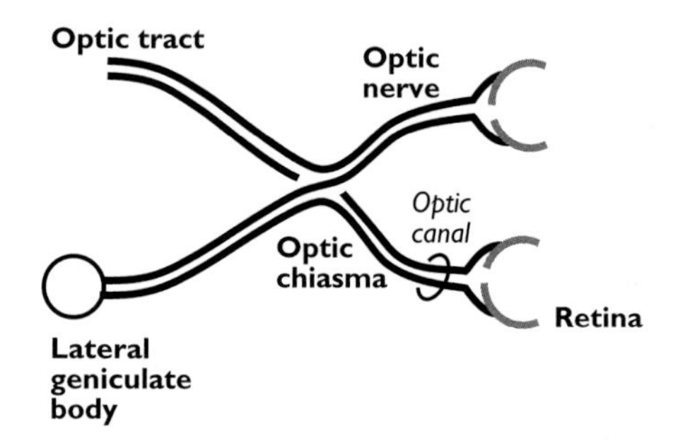

Optic nerve (II)

OLFACTORY NERVE (I)

From: **Olfactory epithelium**
To: **Olfactory cortex**
Contains: **Special sense (smell)**

The olfactory epithelium lines the superior surface of the superior concha, upper medial nasal septum and inferior surface of the cribriform plate of the ethmoid bone. The fibres of the olfactory cells run in the submucosa to pass through the cribriform plate of the ethmoid bone where they synapse in the olfactory bulb which lies on its superior surface. The bulb leads posteriorly to the olfactory tract which lies in the anterior cranial fossa on the inferior surface of the frontal lobe and conveys fibres to the anterior olfactory nucleus (in the posterior aspect of the olfactory bulb), to the prepiriform cortex, anterior perforating substance and septal areas.

OPTIC NERVE (II)

From: **Retina**
To: **Lateral geniculate body**
Contains: **Special sense (sight)**

The ganglion cells of the retina pass fibres out of the globe of the eye via the optic disc to enter the optic N which passes through the orbit within the dural sheath and within the cone of muscles. The nerve passes through the optic canal in the body of the sphenoid bone into the middle cranial fossa where it lies medial to the anterior clinoid process. The ophthalmic artery lies inferior to it in the canal and runs forwards to pierce the dura around the nerve inferomedially about 1 cm behind the eyeball. The nerve continues posteriorly at first lateral to, then superior to, the sella turcica where it forms the optic chiasma. Fibres from both eyes are distributed to each optic tract with medial retinal fibres (temporal visual fields) crossing to the opposite side. Each tract passes from the posterolateral angle of the chiasma, lying lateral to the pituitary infundibulum, to run lateral to the cerebral peduncle and medial to the uncus of the temporal lobe to reach the lateral geniculate body.

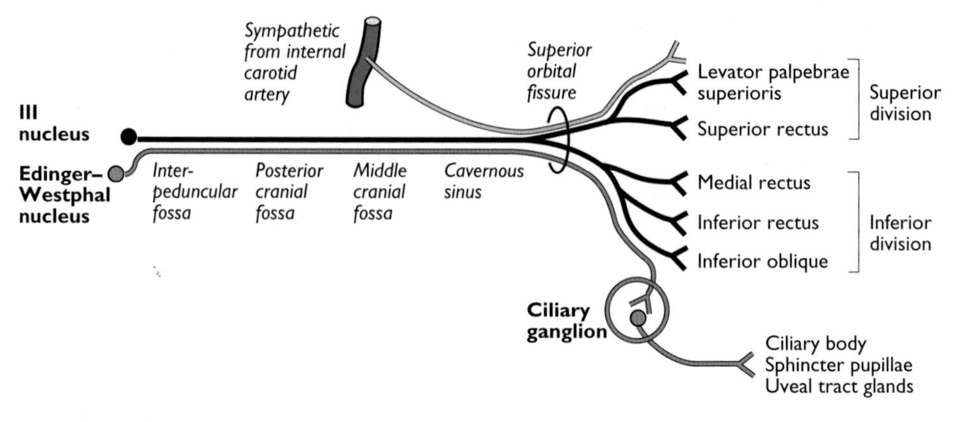

Oculomotor nerve (III)

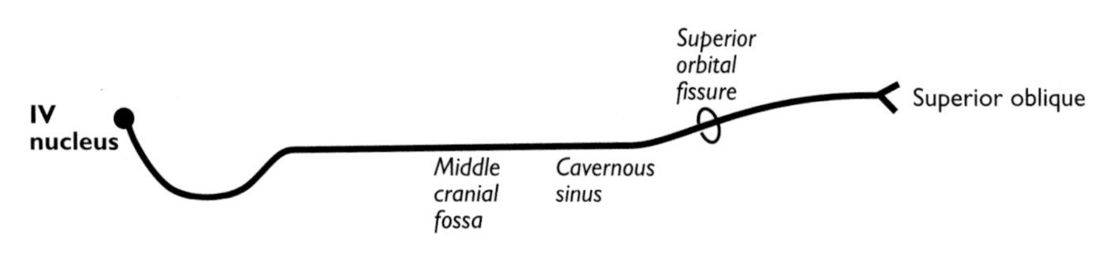

Trochlear nerve (IV)

OCULOMOTOR NERVE (III)

From: **Oculomotor nucleus (somatic motor) and Edinger–Westphal nucleus (general visceral motor), ventral to cranial part of aqueduct in midbrain**
To: **Terminal brs**
Contains: **Somatic motor & general visceral motor**

This nerve emerges medial to the cerebral peduncle in the interpeduncular fossa to reach the middle cranial fossa. It runs forward in close lateral relation to the posterior communicating artery below the margin of the tentorium cerebelli. It pierces the dura lateral to the posterior clinoid process to enter the roof of the cavernous sinus lying initially high in its lateral wall. It descends, passing medially over the trochlear N and nasociliary branch of the ophthalmic division of the trigeminal N. It then enters the orbit through the superior orbital fissure within the tendinous ring having divided into superior and inferior divisions at the anterior end of the cavernous sinus. The superior division runs lateral to the optic N on the inferior surface of the superior rectus, passing through this muscle to terminate in levator palpebrae superioris. This division carries sympathetic supply to this muscle from the internal carotid artery in the cavernous sinus. The inferior division divides into terminal branches shortly after passing through the tendinous ring, the nerve to inferior oblique carrying the general visceral motor fibres (parasympathetic) to the ciliary ganglion. This lies posteriorly in the orbit inferolateral to the optic N.

TROCHLEAR NERVE (IV)

From: **Trochlear nucleus in floor of aqueduct in dorsal midbrain, level with upper part of inferior colliculus**
To: **Terminal brs**
Contains: **Somatic motor**

The fibres decussate within the substance of the midbrain to appear on the opposite side. The nerve emerges dorsally and passes lateral to the superior cerebellar peduncle then around the lateral aspect of the midbrain in the middle cranial fossa to lie just above the superior border of the pons. It runs below the edge of the tentorium cerebelli between the posterior cerebral and the superior cerebellar arteries. It enters the roof of the cavernous sinus then onto its lateral wall where it is crossed medially by the oculomotor N from above down before entering the orbit through the superior orbital fissure superolateral to the tendinous ring. It runs medially above levator palpebrae superioris to terminate as it pierces superior oblique.

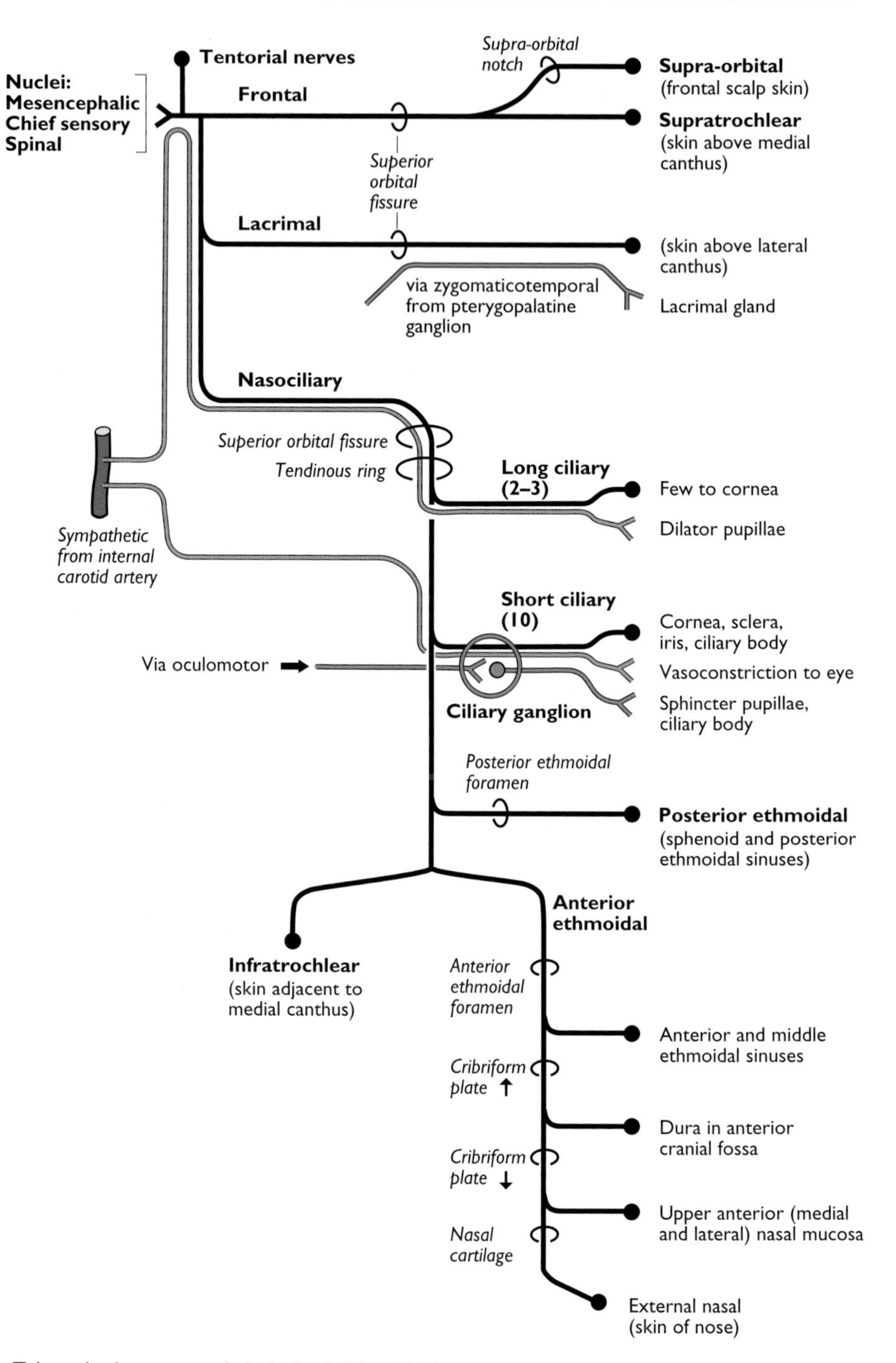

Trigeminal nerve—ophthalmic division (Va)

TRIGEMINAL NERVE—OPHTHALMIC DIVISION (Va)

From: **Terminal nuclei are chief sensory (touch), mesencephalic (proprioception) and spinal (pain & temperature). They lie in pons, midbrain & medulla/upper cervical cord respectively**
To: **Terminal brs**
Contains: **Somatic sensory**

The sensory root of the trigeminal N emerges from the ventral surface of the upper pons to enter the middle cranial fossa from where it passes to the trigeminal ganglion which lies in Meckel's cave, a prolongation of dura at the apex of the petrous temporal bone. The ophthalmic division leaves the trigeminal ganglion and runs forward in the lateral wall of the cavernous sinus below the trochlear N and is crossed medially by the oculomotor N. It divides into three terminal branches which pass through the superior orbital fissure separately. (Note: the mesencephalic nucleus is unusual in that it receives primary neurones that do not have the cell bodies in the ganglion but in the nucleus itself.)

Frontal N. Runs superolateral to the tendinous ring into the orbit where it continues forwards and medially above levator palpebrae superioris, to divide into terminal branches which leave the orbit over its superior margin through similarly named notches.

Lacrimal N. Runs lateral to the tendinous ring into the orbit. It passes laterally, close to the periosteum of the orbital plate of the frontal bone, to supply the lacrimal gland and adjoining conjunctiva. It then leaves the orbit over its superolateral margin to supply the lateral upper eyelid. In its course it carries parasympathetic fibres from the zygomatico-temporal branch of the maxillary N (Vb) to the lacrimal gland and sympathetic fibres from the nearby lacrimal artery.

Nasociliary N. Runs within the tendinous ring between superior and inferior division of the oculomotor N, crossing superior to the optic N to lie over the medial rectus. It leaves the muscular cone giving terminal branches before running through the anterior ethmoidal foramen in the ethmoid bone on the medial orbital wall as the anterior ethmoidal N. It traverses the anterior ethmoid sinus to run through its roof and onto the superior surface of the cribriform plate beneath the dura of the anterior cranial fossa. It passes through the plate again lateral to the crista galli onto the medial wall of the nose, first on the perpendicular plate of the ethmoid and then on the inner surface of the nasal bone. It passes into the skin of the nose beneath the inferior margin of the nasal bone as its terminal branch—the external nasal N.

Short ciliary Ns. Each of these 8–10 nerves emerges from the ciliary ganglion with a sensory component from the nasociliary N, postganglionic parasympathetic fibres from the oculomotor N and sympathetic fibres which pass directly from the internal carotid artery in the cavernous sinus.

Long ciliary Ns. Each of the 2–3 nerves has sensory fibres from the nasociliary N and sympathetic fibres for the dilator pupillae that are carried by the nasociliary N from the internal carotid artery in the cavernous sinus.

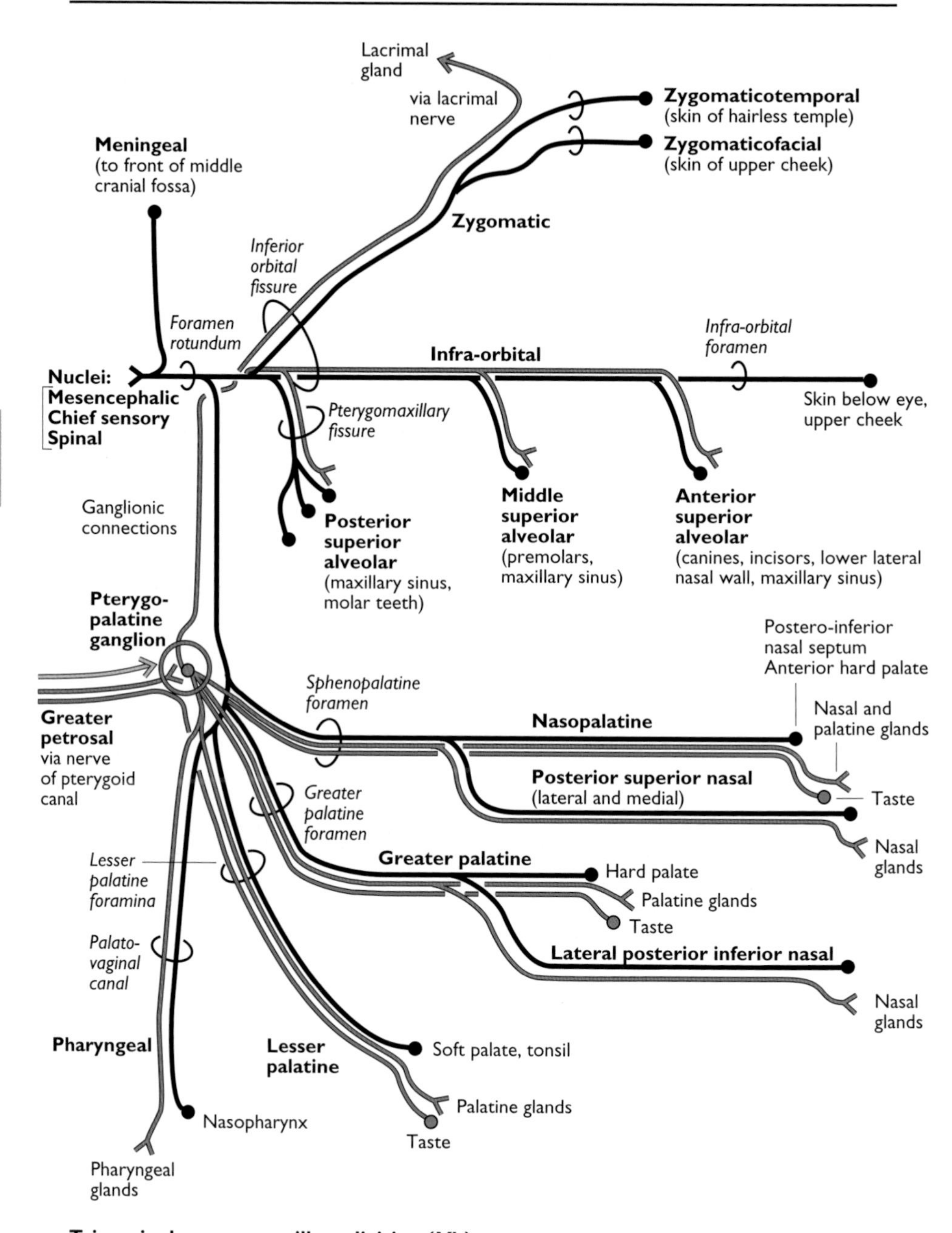

Trigeminal nerve—maxillary division (Vb)

TRIGEMINAL NERVE—MAXILLARY DIVISION (Vb)

From: **Terminal nuclei are chief sensory (touch), mesencephalic (proprioception) and spinal (pain & temperature). They lie in pons, midbrain & medulla/upper cervical cord respectively**
To: **Terminal brs**
Contains: **Somatic sensory**

(See ophthalmic division for course to the trigeminal ganglion.) The nerve leaves the ganglion to run low down in the lateral wall of the cavernous sinus before passing onto the floor of the middle cranial fossa and it then exits through the foramen rotundum in the greater wing of the sphenoid bone. It runs into the upper pterygopalatine fossa, giving branches via the pterygopalatine ganglion before passing into the orbit via the inferior orbital fissure to become the infra-orbital N. The pterygopalatine ganglion is suspended from the maxillary N by one or two roots and receives parasympathetic fibres from the greater petrosal N in the pterygopalatine fossa which are distributed with the terminal branches as shown.

Infra-orbital N. Passes laterally across the posterior aspect of the palatine bone and maxilla to pass through the inferior orbital fissure and to run into the infra-orbital groove in the orbital surface of the maxilla. It terminates in branches in the skin of the face having passed through the infra-orbital foramen which is the completed continuation of the infra-orbital groove.

Zygomatic N. Passes superiorly out of the pterygopalatine fossa through the inferior orbital fissure to run in the lateral orbit outside the cone of muscles. It terminates on the lateral orbital wall as facial and temporal branches which pass through unnamed canals in the zygomatic bone to reach the skin over the zygomatic bone and hairless temple respectively.

Nasopalatine (previously sphenopalatine) N. Passes through the sphenopalatine foramen to enter the posterior upper nasal cavity to end as terminal branches.

Pharyngeal N. Branches of this nerve leave the pterygopalatine ganglion and pass through the palatovaginal canal to supply sensation and secretomotor fibres to the nasopharynx.

A few fine branches from the ganglion also supply the sphenoid and ethmoid sinuses with sensation and secretomotor fibres.

Note: Other branches pass through foramina as indicated. Sympathetics have been omitted from the distribution of the pterygopalatine ganglion to avoid complicating the diagram.

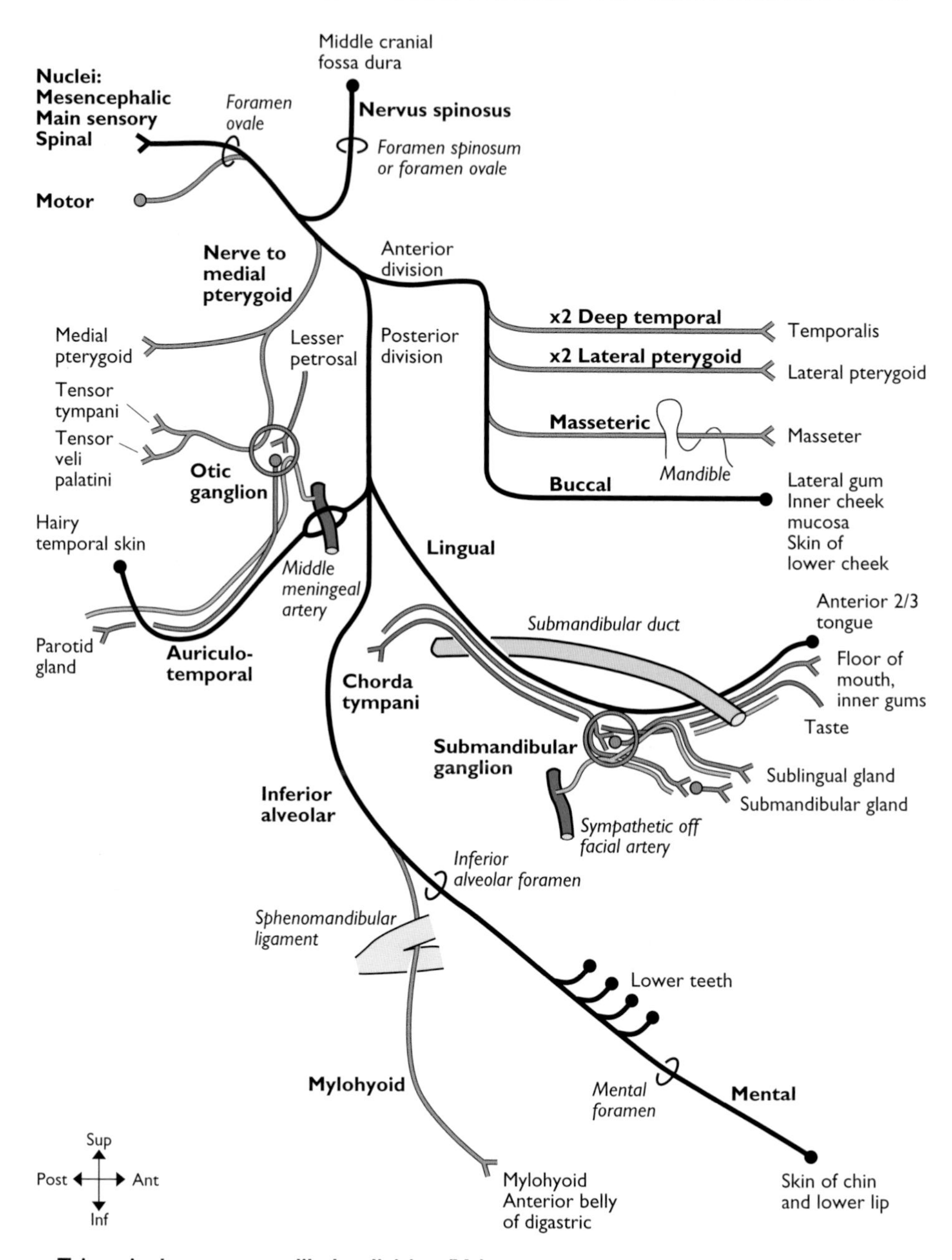

Trigeminal nerve—mandibular division (Vc)

TRIGEMINAL NERVE—MANDIBULAR DIVISION (Vc)

From: **Terminal nuclei are chief sensory (touch), mesencephalic (proprioception) and spinal (pain & temperature). They lie in pons, midbrain & medulla/upper cervical cord respectively. Motor nucleus (branchial muscles) is in upper pons**
To: **Terminal brs**
Contains: **Somatic sensory & special visceral motor**

(See ophthalmic division for course to the trigeminal ganglion.) The smaller motor root leaves the ventral pons anteromedial to the sensory root. The sensory root leaves the ganglion from its lateral part and passes, after a short course over the greater wing of the sphenoid bone, through the foramen ovale. The motor root passes under the ganglion and unites with the sensory root just beyond the foramen ovale. The nerve that is so formed passes into the infratemporal fossa between tensor veli palatini and lateral pterygoid. It has a short course of 3–4mm before dividing into anterior and posterior divisions which provide terminal branches.

Anterior division

Deep temporal Ns. Usually two, run above the lateral pterygoid, over the infratemporal crest and on the squamous temporal and greater wing of the sphenoid bones deep to temporalis to supply it. They lie with their associated vessels deeply in the temporal fossa.

Lateral pterygoid Ns. Pierce the muscle directly to supply it.

Masseter N. Runs laterally over the lateral pterygoid and over the mandibular notch to pierce the deep surface of the muscle to supply it and the temporomandibular joint.

Buccal N. Runs forward over lateral pterygoid and lies deep to temporalis, the mandible and masseter. It runs through buccinator to supply a small area of overlying skin. The nerve is sensory only.

Posterior division

Auriculotemporal N. Passes posteriorly, briefly dividing to encircle the middle meningeal artery, before running between the neck of the mandible and the spheno-mandibular ligament. It winds around the neck of the mandible to pass laterally then superiorly lying between the temporo-mandibular joint and the external auditory meatus deep to the parotid gland. It terminates at the upper border of the gland as branches. The nerve receives general visceral motor fibres (parasympathetic) from the lesser petrosal N via the otic ganglion which lies suspended from the nerve to tensor tympani near the foramen ovale.

Lingual N. Passes forward and inferiorly to lie between lateral pterygoid and tensor veli palatini, then between medial pterygoid and the ramus of the mandible. It lies just beneath the mucous membrane of the mouth posteromedial to the third molar tooth. It passes lateral to styloglossus and hyoglossus and runs at first lateral, then inferior and then medial to the submandibular duct. It terminates over the lateral aspect of the anterior two-thirds of the tongue. It is joined 2cm anterior and inferior to the foramen ovale by general visceral motor and special visceral sensory fibres of the chorda tympani which relay in the submandibular ganglion. This ganglion which is suspended from the nerve by two roots lies on hyoglossus above the submandibular gland.

Inferior alveolar N. Passes deep to lateral pterygoid to lie between the spheno-mandibular ligament and the ramus of the mandible before entering the mandible via the inferior alveolar foramen. It terminates as cutaneous branches reappearing through the mental foramen in the anterior body of the mandible. The N to mylohyoid arises just before it enters the inferior alveolar foramen and pierces the sphenomandibular ligament to run in a groove on the medial surface of the body of the mandible below mylohyoid.

5

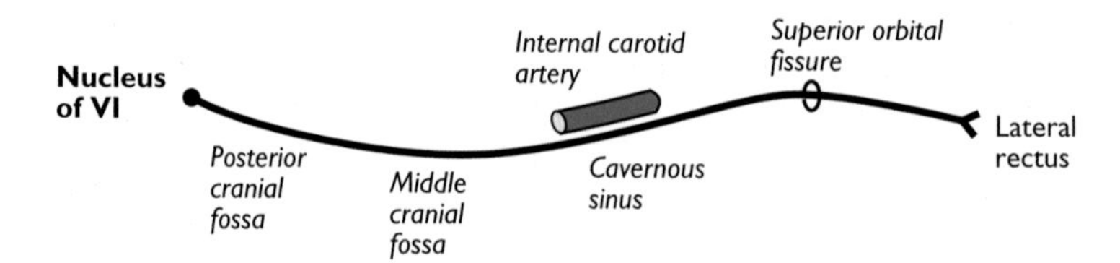

Abducent nerve (VI)

ABDUCENT NERVE (VI)

From: **Abducent nucleus in sup part of floor of 4th ventricle in lower pons**
To: **Terminal brs**
Contains: **Somatic motor**

The fibres leave the pons at its lower border above the pyramid of the medulla. It traverses the pontine basal cistern running forwards and superiorly to pierce the dura over the clivus inferolateral to the dorsum sellae. It arches forward directly over the ridge of the petrous temporal bone passing through the medial wall of the inferior petrosal sinus under the petroclinoid ligament and runs into the cavernous sinus. Here it lies directly lateral to the internal carotid artery before passing into the orbit through the superior orbital fissure within the tendinous ring. It passes forward and laterally to sink into the medial surface of the lateral rectus muscle.

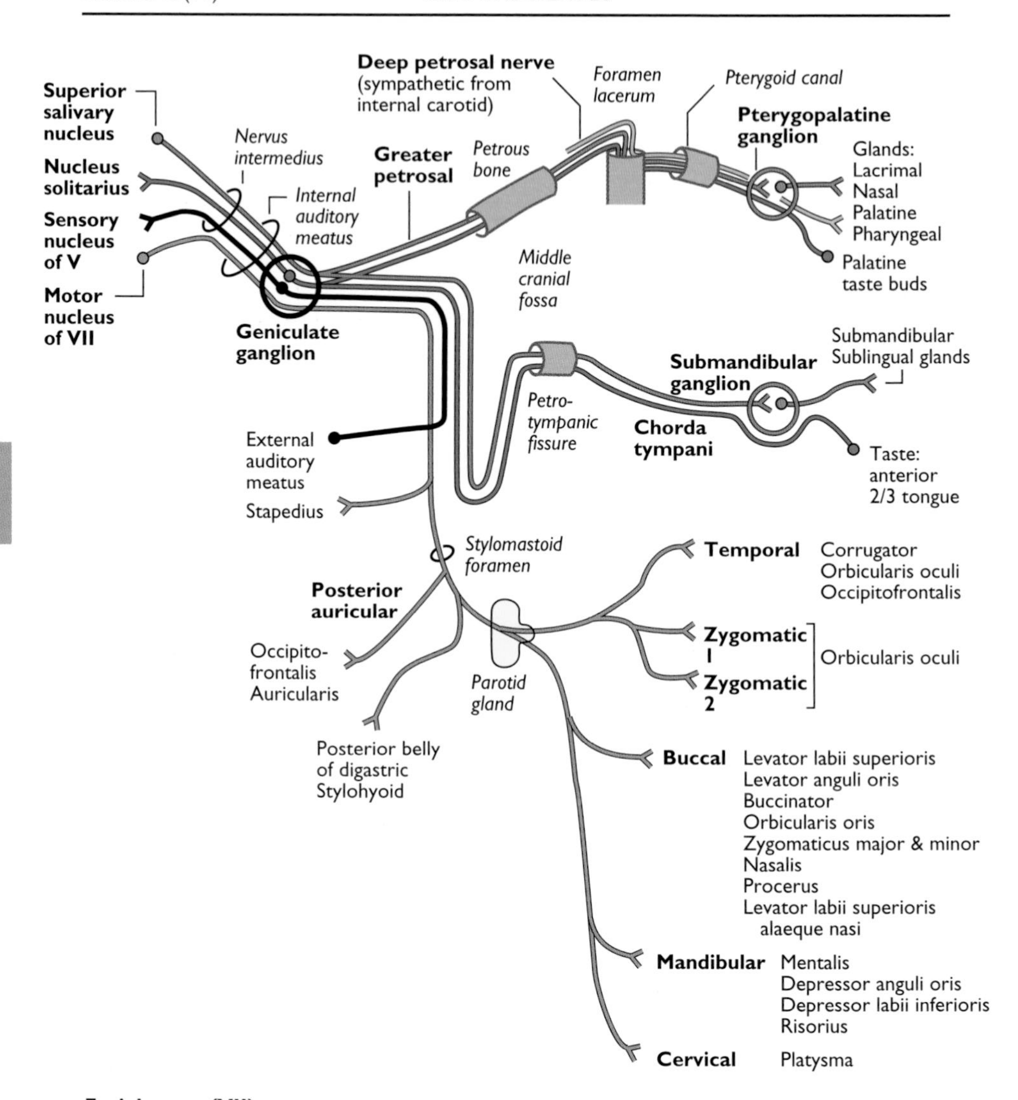

Facial nerve (VII)

FACIAL NERVE (VII)

From: **Facial motor nucleus deep to reticular formation in lower pons. Sup salivary nucleus (general visceral motor) distal to motor nucleus. Gustatory nucleus (taste) in superior end of nucleus solitarius in medulla. Sensory nucleus of V (see trigeminal N)**
To: **Terminal brs**
Contains: **Somatic sensory, special visceral sensory & motor, general visceral motor**

It leaves the pons at the cerebellopontine angle medial to the vestibulocochlear N (VIII) as two nerve roots—the facial motor root and the nervus intermedius. The nervus intermedius contains special visceral sensory, general visceral motor and somatic sensory fibres which connect with gustatory, superior salivary and sensory trigeminal nuclei respectively. These two roots pass across the subarachnoid space together to enter the internal auditory meatus and pass laterally along it to reach and enter the facial canal. The two roots then unite and pass laterally onto the medial wall of the middle ear before turning 90 degrees posteriorly at the geniculate ganglion. It continues posteriorly running above the promontory and oval window and below the lateral semicircular canal. Finally the nerve turns 90 degrees inferiorly to run down in the medial wall of the aditus of the mastoid antrum. It leaves the middle ear via the stylomastoid foramen to pass between the mastoid process and the tympanic ring before passing between the deep and superficial portions of the parotid gland. Within the gland it lies superficial to the styloid process, retromandibular vein and external carotid artery before dividing into terminal branches which leave just medial to the anterior border of the gland.

Greater petrosal N. Arises from the main nerve at the geniculate ganglion and passes medially through the petrous temporal bone to lie in a groove on its anterior surface beneath the temporal lobe and dura of the middle cranial fossa. It runs beneath the trigeminal ganglion and then passes anteromedially but lying anterolateral to the internal carotid artery as this vessel emerges across the upper end of foramen lacerum. It is joined by the deep petrosal N (sympathetic) off the artery and this combined nerve (the N of the pterygoid canal or Vidian's N) passes into the same named canal which commences a short way into the anterior wall of the foramen lacerum at the root of the pterygoid process of the sphenoid. This canal exits into the pterygopalatine fossa where the nerve enters the posterior aspect of the pterygopalatine ganglion and distributes its fibres with branches of the maxillary division (Vb) of the trigeminal N.

Chorda tympani. Arises from the facial nerve in the facial canal during its descent from the medial wall of the middle ear. It runs back into the middle ear on the posterior wall before passing anteriorly between the two layers of the flaccid part of the tympanic membrane and over the handle of the malleus. It leaves the middle ear by passing into the petrous temporal bone and emerges via the petrotympanic fissure to pass into the infratemporal fossa medial to the spine of the sphenoid bone which it grooves. It runs antero-inferiorly, deep to lateral pterygoid, to join the lingual branch of the mandibular division (Vc) 2 cm below the skull.

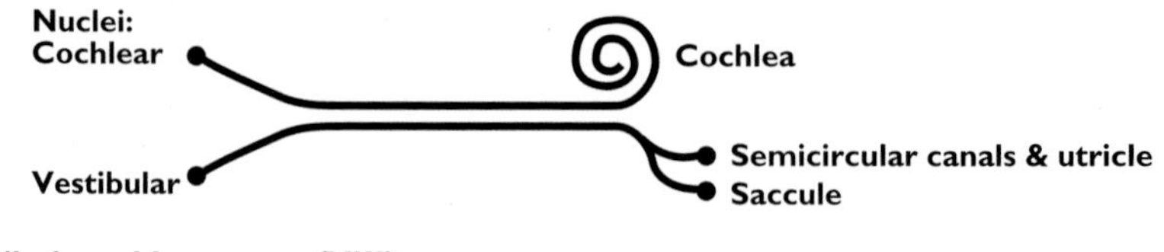

Vestibulocochlear nerve (VIII)

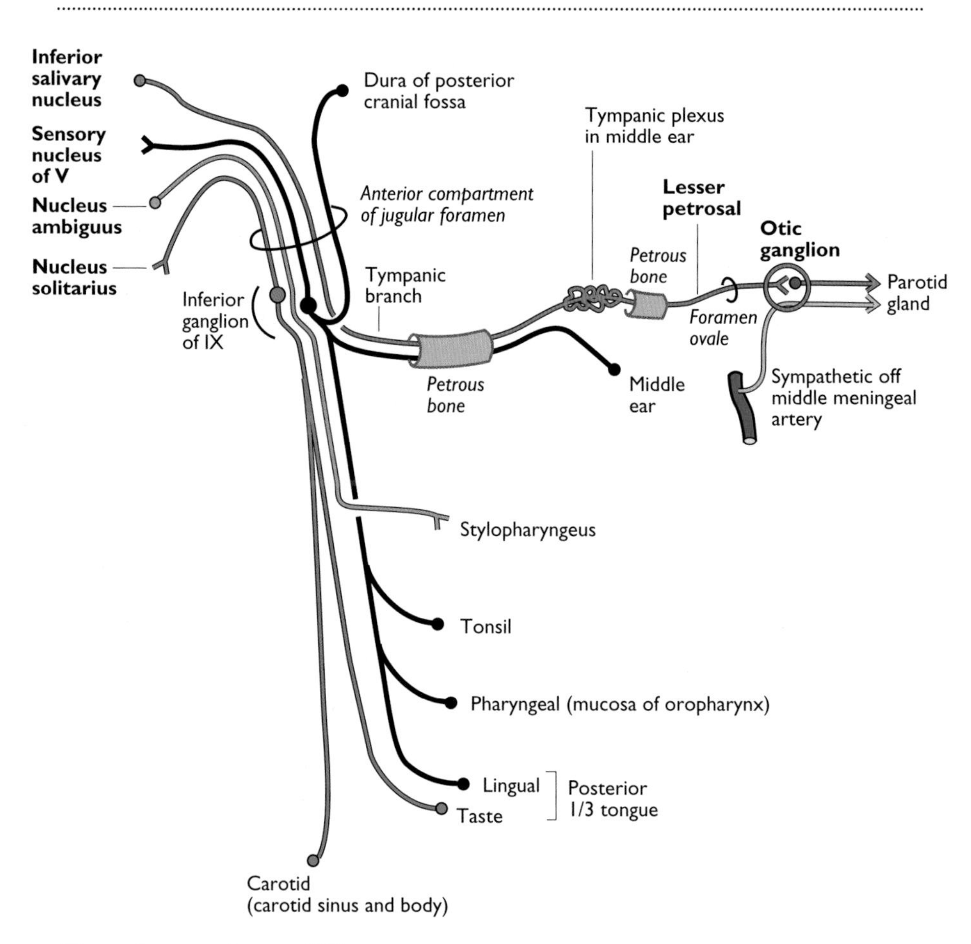

Glossopharyngeal nerve (IX)

VESTIBULOCOCHLEAR NERVE (VIII)

From: **4 Vestibular & 2 cochlear nuclei in floor of 4th ventricle in pons**
To: **Inner ear**
Contains: **Special sense (hearing, balance)**

It emerges at the cerebellopontine angle as a single nerve and traverses the subarachnoid space to enter the internal auditory meatus where the cochlear element separates and pierces the temporal bone in its antero-inferior quadrant. The vestibular element divides into upper and lower divisions to pierce the temporal bone in its postero-superior and postero-inferior quadrants. The cochlear N runs in the cochlear modiolus to end in terminal connections. The upper vestibular division runs to supply the semicircular canals and the utricle, the lower division the saccule.

GLOSSOPHARYNGEAL NERVE (IX)

From: **Sensory nucleus of V (common sensation—see trigeminal N). Nucleus solitarius (taste—medulla). Nucleus ambiguus (motor to branchial muscle—medulla). Inf salivary nucleus (secretomotor—lower pons)**
To: **Terminal brs**
Contains: **Somatic sensory, general & special visceral motor, special visceral sensory**

The fibres leave the medulla as three or four rootlets lying posterior to the olive. They rapidly fuse into one nerve which passes anterolaterally into the anterior compartment of the jugular foramen (between petrous temporal and occipital bones). It passes lateral to the inferior petrosal sinus which separates it from the vagus and accessory Ns and runs anteriorly out of the compartment. It forms the glossopharyngeal ganglia below the compartment as it passes between internal jugular vein and internal carotid artery. It passes inferolaterally looping around the upper border of stylopharyngeus, runs deep to hyoglossus and terminates in lingual and pharyngeal branches.

Carotid N. Arises just below the ganglia and runs down closely adherent to the internal carotid artery within the carotid sheath to reach the carotid sinus and carotid body.

Lesser petrosal N. The tympanic branch (Jacobson's N) of the glossopharyngeal N arises just below the ganglia to pass via the petrous temporal bone into the middle ear. It mingles with parasympathetic fibres of the facial N (VII) and sympathetics from the internal carotid artery over the promontory on the medial wall as the tympanic plexus before forming the lesser petrosal N. This nerve then leaves the middle ear via the medial side of the roof and passes through the petrous temporal bone into the middle cranial fossa. Here it lies beneath the dura to run forward before passing through the foramen ovale and synapsing below in the otic ganglion in the infratemporal fossa. From here it is distributed with the auriculo-temporal branch of the mandibular division (Vc) to the parotid gland.

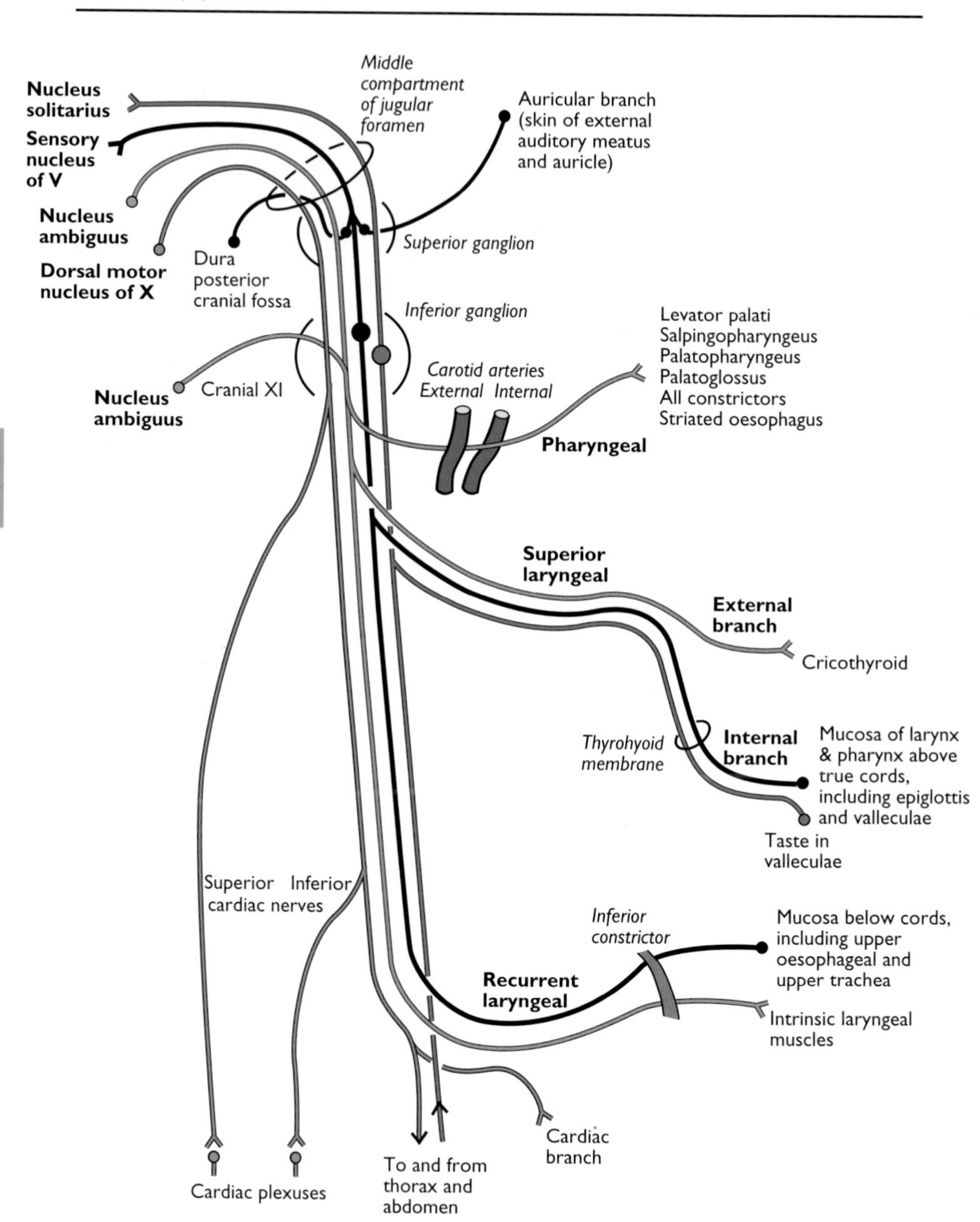

Vagus nerve (X)

VAGUS NERVE (X)

From: **Dorsal motor nucleus of vagus (general visceral motor–lower medulla). Nucleus ambiguus (branchial motor–medulla). Nucleus solitarius (taste & general visceral sensory–medulla). Sensory nucleus of V (common sensation–see trigeminal N)**
To: **Terminal brs**
Contains: **Somatic sensory, general & special visceral sensory, general & special visceral motor**

The fibres emerge from the medulla as a series of rootlets posterior to the olive between the glossopharyngeal and cranial accessory rootlets. These form into a single nerve that passes into the middle compartment of the jugular foramen. Below the foramen it forms superior and inferior ganglia before being joined by the cranial part of the accessory N. It passes vertically down within the carotid sheath closely related to the internal carotid artery and lying between it and the internal jugular vein.

Pharyngeal branch. Passes from the vagus at the inferior ganglion running between internal and external carotid arteries to reach the lateral wall of the pharynx. These fibres are mostly from the cranial part of the accessory N.

Superior laryngeal N. Passes from the inferior ganglion running steeply down anteriorly, lying posterior and then medial to the internal carotid artery. It pierces the carotid sheath to run on the wall of the pharynx to the level of the greater cornu of the hyoid bone where it divides.

Internal branch. Runs down anteriorly onto the thyrohyoid membrane which it pierces at the level of the vallecula, and is then distributed as terminal branches to the mucous membrane of the vallecula and larynx down to the vocal folds.

External branch. Runs down over the inferior constrictor accompanied by the superior thyroid artery to reach the cricothyroid muscle.

Recurrent laryngeal N. In the neck the two sides follow the same course ascending in the tracheo-oesophageal groove. As the nerve passes medial to the lateral lobe of the thyroid gland it is intimately related to the inferior thyroid artery. It passes beneath the inferior border of cricopharyngeus (inferior constrictor) to terminate within the submucosa of the larynx. The nerve on the right originates from the vagus anterior to the subclavian artery around which it hooks posteriorly before running medially to ascend in the tracheo-oesophageal groove. The nerve on the left originates from the vagus inferolateral to the arch of the aorta passing inferior to the arch and posterior to the ligamentum arteriosum. It runs to the right of the arch as it passes posteriorly over the left side of the trachea to reach the tracheo-oesophageal groove.

Cardiac Ns. The upper branch arises below the inferior ganglion and the lower branch arises in the root of the neck. On the right they pass down anterior to the brachiocephalic artery and on the left over the aortic arch to terminate in the cardiac plexuses.

Note: General visceral motor (parasympathetic) and general visceral sensory fibres in the vagus to and from the thorax and abdomen arise/end in the dorsal motor nucleus of vagus and nucleus solitarius respectively.

All branchial motor (special visceral motor) fibres in the vagus which supply muscles arise in the nucleus ambiguus but the majority, if not all, are supplied to the vagus by the cranial accessory (XI).

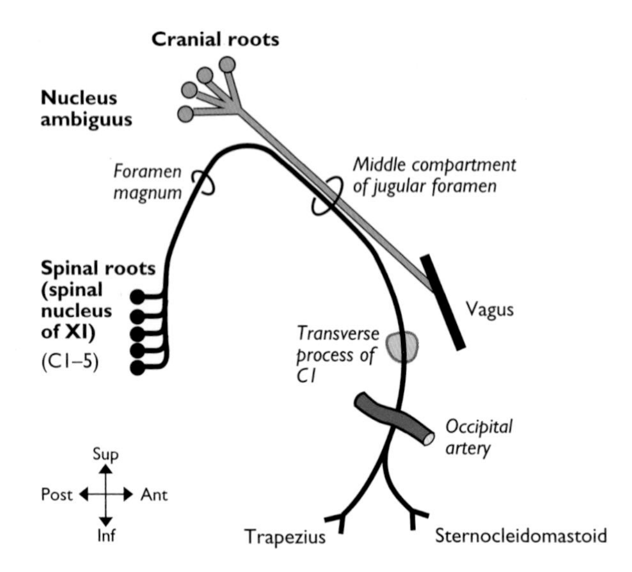

Accessory nerve (XI)

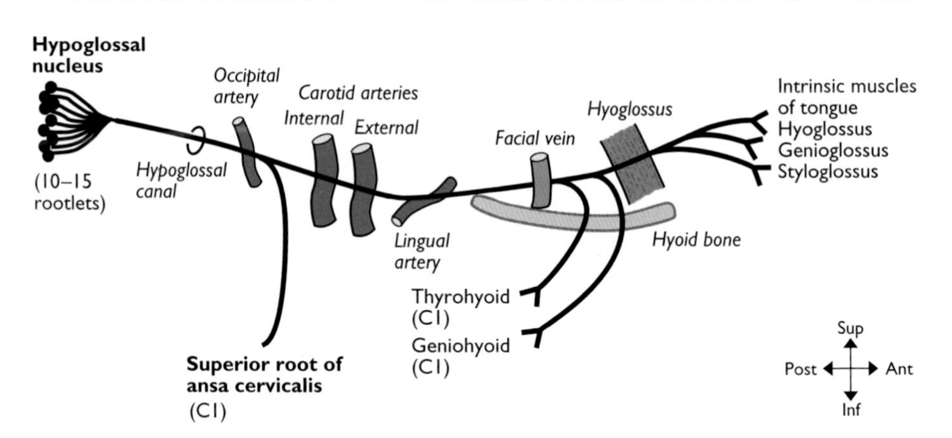

Hypoglossal nerve (XII)

ACCESSORY NERVE (XI)

From: **Cranial root from nucleus ambiguus (branchial motor—medulla). Spinal root from spinal nuclei (C1–C5)**
To: **Terminal brs**
Contains: **Somatic motor (spinal), special visceral motor (cranial)**

The fibres of the cranial root emerge from the medulla as four to six rootlets posterior to the olive immediately below those of the vagus to fuse into a single nerve. They are joined by the spinal root as it ascends via the foramen magnum (see cervical plexus, pp. 112–113). The nerve passes out of the posterior cranial fossa through the middle compartment of the jugular foramen posterior to the vagus and anterior to the internal jugular vein. Inferior to the foramen the cranial element passes inferomedially to fuse with the vagus to which it adds its complement of special visceral motor fibres.

Spinal root. Passes posterolaterally, usually posterior to the internal jugular vein, over the lateral mass of the atlas (C1) and deep to the occipital artery to enter the deep surface of sternocleidomastoid. It traverses the posterior triangle of the neck from one-third of the way down the posterior border of sternocleidomastoid to one-third of the way up the anterior border of trapezius where it terminates.

HYPOGLOSSAL NERVE (XII)

From: **Hypoglossal nucleus in floor of 4th ventricle in medulla**
To: **Terminal brs**
Contains: **Somatic motor**

Its fibres pass out of the anterolateral surface of the medulla between the olive and pyramid as a series of 10–15 rootlets. These fuse to form two roots which pass posterior to the vertebral artery as they run into the hypoglossal canal where they themselves fuse. The nerve runs out of the canal anteriorly, lateral to the occipital, internal carotid, external carotid and lingual arteries before passing over the apex of the greater cornu of the hyoid bone. It then runs anteriorly, looping lateral to hyoglossus, deep to mylohyoid, to end in terminal branches beneath the submandibular gland.

Note: C1 fibres join the hypoglossal N (pp. 112–113). This is not shown in this diagram but they are seen leaving it.

6: PERIPHERAL NERVES

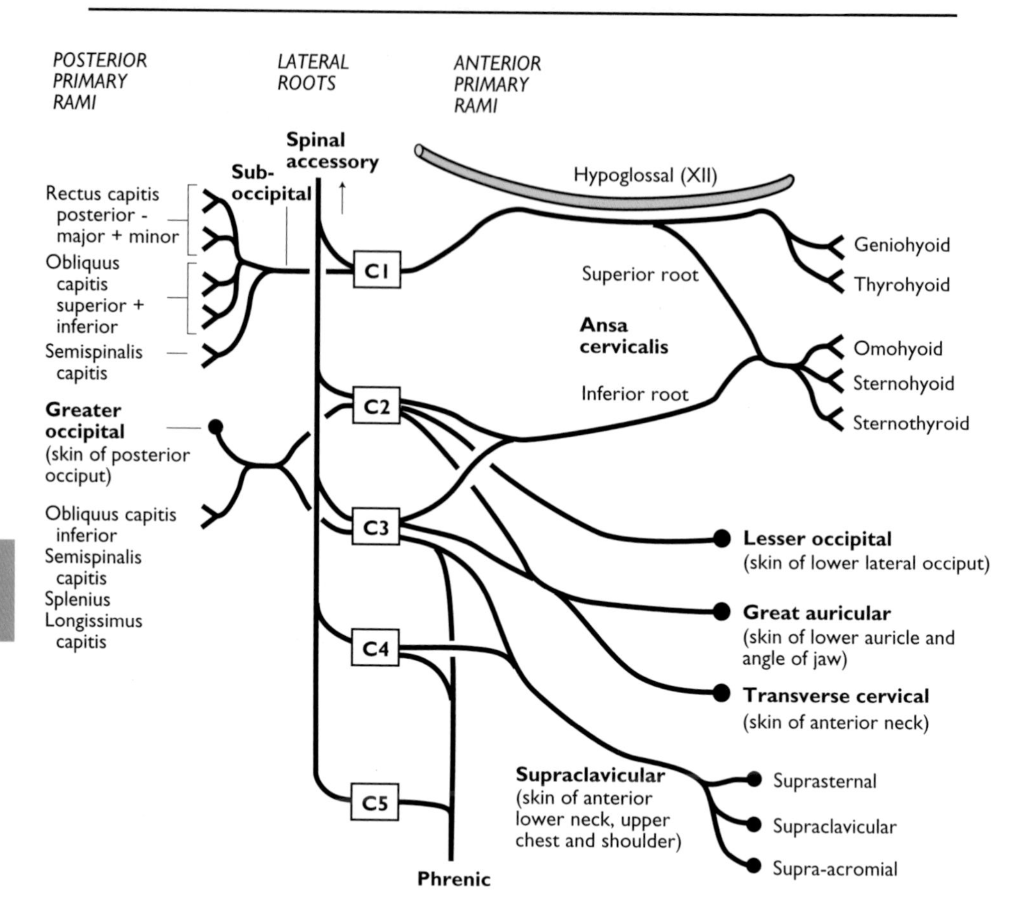

Cervical plexus (C1,2,3,4,5)

CERVICAL PLEXUS (C1,2,3,4,5)
From: C1,2,3,4,5 Ns
To: Ns as shown

It arises mostly from the anterior primary rami deep between scalenus medius and scalenus anterior at the level of C1–C4 vertebrae and is covered by prevertebral fascia lying deep to sternocleidomastoid. The cutaneous branches pierce the prevertebral fascia and run into the posterior triangle of the neck where they pierce the investing layer of the deep cervical fascia to terminate in subcutaneous Ns.

Ansa cervicalis (C1–C3). Superior root (anterior primary rami C1)—passes directly to the hypoglossal N (XII) between rectus capitis anterior and lateralis. It leaves the hypoglossal N lateral to the occipital artery and runs anterior to the internal and common carotid arteries where it joins the inferior root. Inferior root (anterior primary rami C2,3)—passes laterally around the internal jugular vein having pierced the prevertebral fascia at the level of C2/3. It runs forwards and anteriorly as a long loop to meet with the superior root anterior to the common carotid artery.

Suboccipital N (posterior primary ramus of C1). Emerges through the dura to run beneath the vertebral artery closely applied to the posterior arch of the atlas (C1). It pierces the posterior atlanto-occipital membrane between obliquus capitis superior and rectus capitis posterior major to terminate in muscular branches in the suboccipital triangle.

Greater occipital N (posterior primary rami of C2, and a small contribution from C3). Emerges from the posterior spinal dura at the intervertebral foramen and passes posteriorly over the transverse process of the axis (C2) below obliquus capitis inferior. It then winds around this muscle to ascend deep to semispinalis piercing it and trapezius near to their insertions into the superior nuchal line. It terminates as cutaneous branches running in the scalp with the occipital artery.

Spinal accessory N (XIs) (lateral roots C1–C5) is formed from the unique lateral roots of C1–C5 and ascends within the subarachnoid space lateral to the cord and posterior to the denticulate ligament, to pass through the foramen magnum posterior to the vertebral artery to meet with the cranial root.

Phrenic N (see pp. 124–125).

Great auricular N (C2,3). Supplies skin over parotid gland and angle of jaw (this is the only area of the face that is not supplied by a cranial nerve), lower lateral auricle below external auditory meatus, and whole of posterior (medial) auricle.

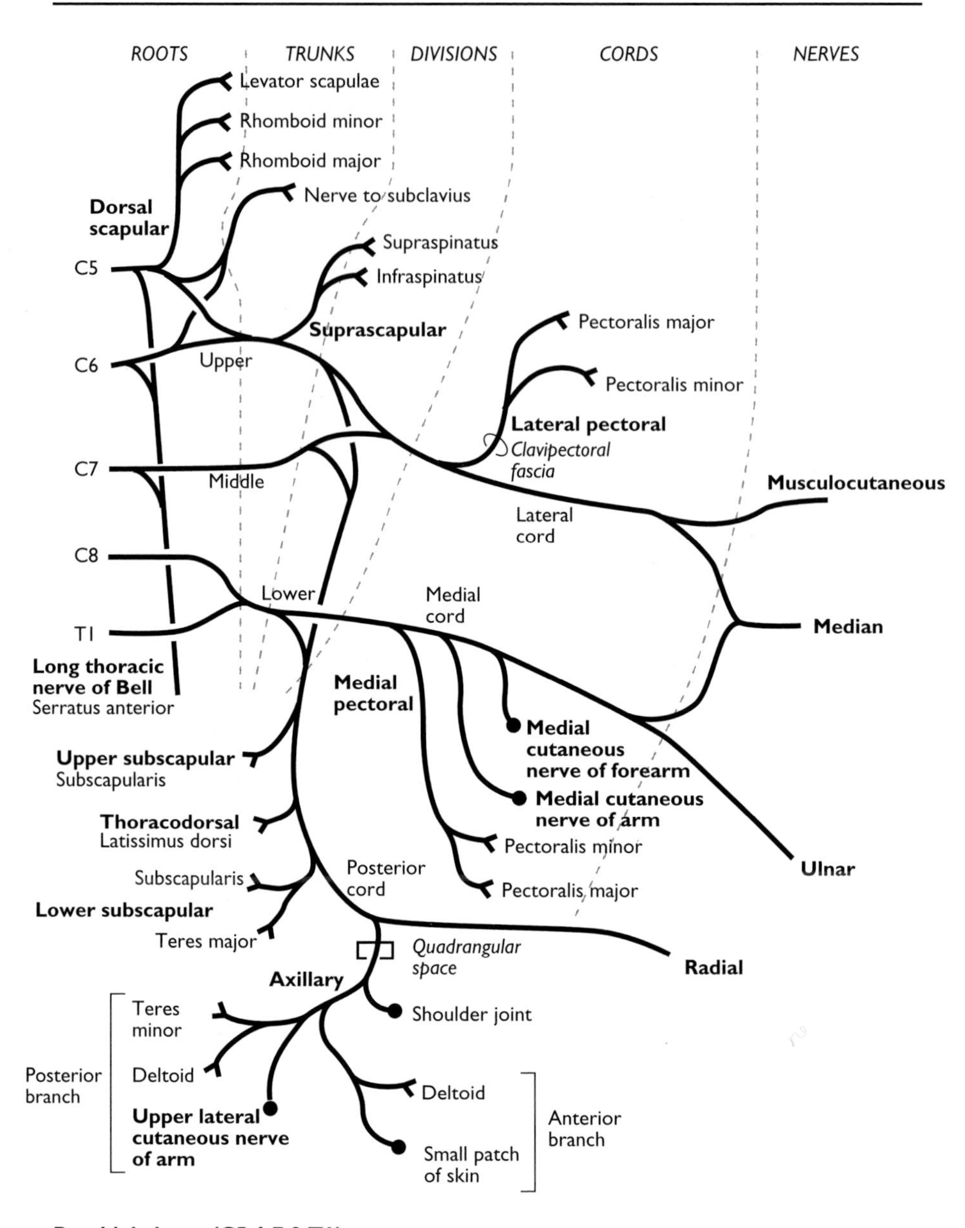

Brachial plexus (C5,6,7,8,T1)

BRACHIAL PLEXUS (C5,6,7,8,T1)
From: **Ant primary rami of C5,6,7,8,T1**
To: **Musculocutaneous, median, ulnar & radial Ns**

It emerges as five roots lying anterior to scalenus medius and posterior to scalenus anterior. The trunks lie in the base of the posterior triangle of the neck, where they are palpable, and pass over the 1st rib posterior to the third part of subclavian artery to descend to lie behind the clavicle. The divisions form behind the middle third of the clavicle lying on the upper fibres of serratus anterior and around the axillary artery, as they form the cords. The cords lie in the axilla related to the second part of the axillary artery lying medial, lateral and posterior as their names indicate and posterior to pectoralis minor. Terminal nerves are formed around the third part of the axillary artery posterior to the lower fibres of pectoralis major.

Axillary N (C5,6)**.** Arises posterior to the third part of axillary artery. It runs posteriorly on subscapularis to pass through the quadrangular space with the posterior circumflex humeral artery. It is intimately related to the medial aspect of the surgical neck of the humerus running laterally to end in anterior and posterior divisions deep to deltoid.

Upper lateral cutaneous N of arm. Is a continuation of the posterior branch of the axillary N and supplies skin over the lower lateral deltoid. Damage to the axillary N (and hence this nerve) gives a 'regimental badge' area of sensory loss.

Thoracodorsal N (C6,7,8)**.** Runs with the subscapular artery down the medial scapular border over teres major and into latissimus dorsi.

Long thoracic N of Bell (C5,6,7)**.** Descends posterior to the trunks of the plexus and the first part of the axillary artery to lie on the lateral aspect of serratus anterior on the medial axillary wall.

Suprascapular N (C5,6)**.** Arises in the posterior triangle of the neck, passes posterolaterally deep to trapezius and omohyoid and runs through the suprascapular notch into the supraspinous fossa. It descends laterally around the scapular spine into the infraspinous fossa.

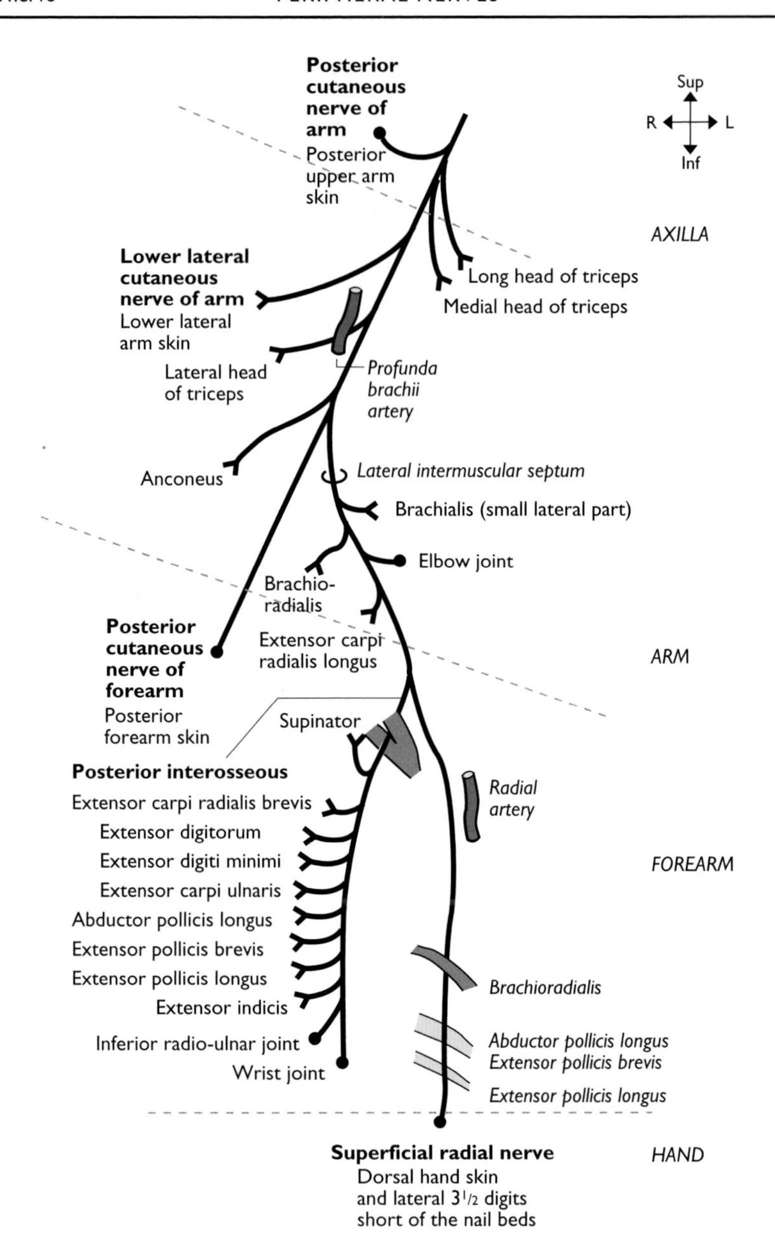

Radial nerve (C5,6,7,8,T1)

RADIAL NERVE (C5,6,7,8,T1)
From: **Post cord of brachial plexus**
To: **Terminal brs**

It arises as the continuation of the posterior cord and descends posterior to the axillary and brachial arteries passing inferior to the tendons of latissimus dorsi and teres major to run with the profunda brachii artery between the long and medial heads of triceps and through the lateral triangular space. It gives off the posterior cutaneous N of arm before leaving the axilla. It then runs over the spiral line of the humerus between medial and lateral heads of triceps giving muscular and cutaneous branches, and pierces the lateral intermuscular septum at the mid point of the humerus to reach the anterior compartment. Here it lies deep to the upper fibres of brachialis then brachioradialis, before entering the lateral cubital fossa. It divides into terminal branches over the lateral epicondyle.

Superficial terminal branch. Runs over supinator, pronator teres and flexor digitorum superficialis and lies under brachioradialis running with the radial artery on its medial aspect from one-third of the way down the forearm. It passes posteriorly, emerging from under the tendon of brachioradialis proximal to the radial styloid and then passes over the tendons of the snuff box where it terminates as cutaneous branches to the dorsum of the hand.

Posterior interosseous N. Passes between the two heads of supinator three fingers' breadth below radial head passing into the posterior compartment where it breaks up into terminal muscular branches in the plane between the deep and superficial muscles in this compartment.

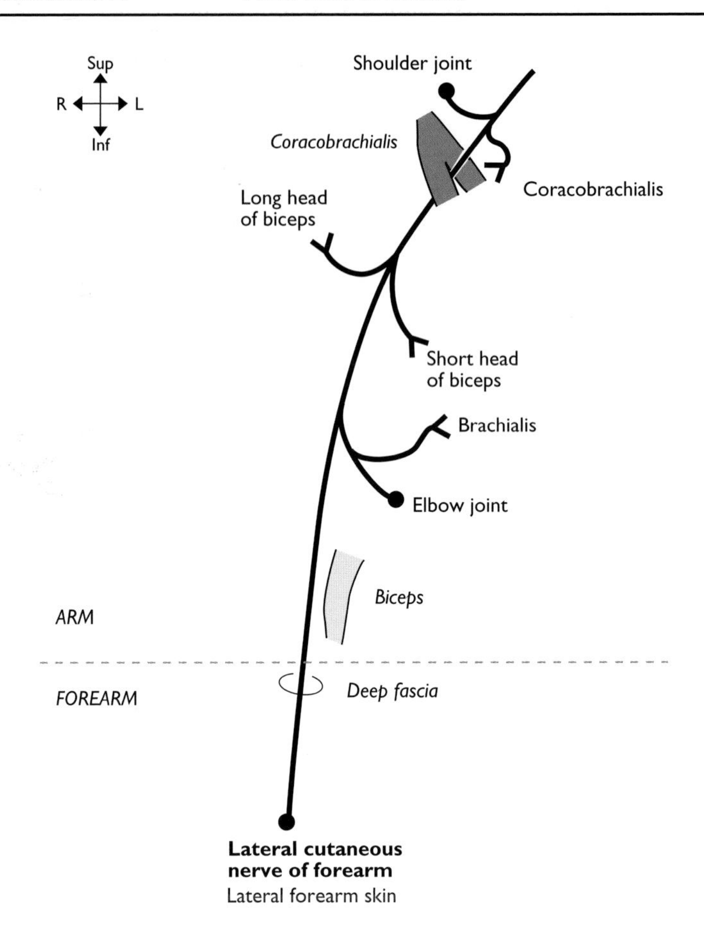

Musculocutaneous nerve (C5,6,7)

MUSCULOCUTANEOUS NERVE (C5,6,7)

From: Lat cord
To: Terminal brs

It arises obliquely behind the lower fibres of pectoralis minor lying lateral to the axillary artery and passes laterally between the two conjoined heads of coracobrachialis. It runs laterally downwards between biceps and brachialis, usually adherent to the deep surface of biceps. The terminal branch is the lateral cutaneous N of forearm.

Lateral cutaneous N of forearm. Emerges lateral to the tendon of biceps in the cubital fossa, piercing the deep fascia just below the elbow and descends over the lateral aspect of the forearm to terminate in the skin over the radial artery at the wrist.

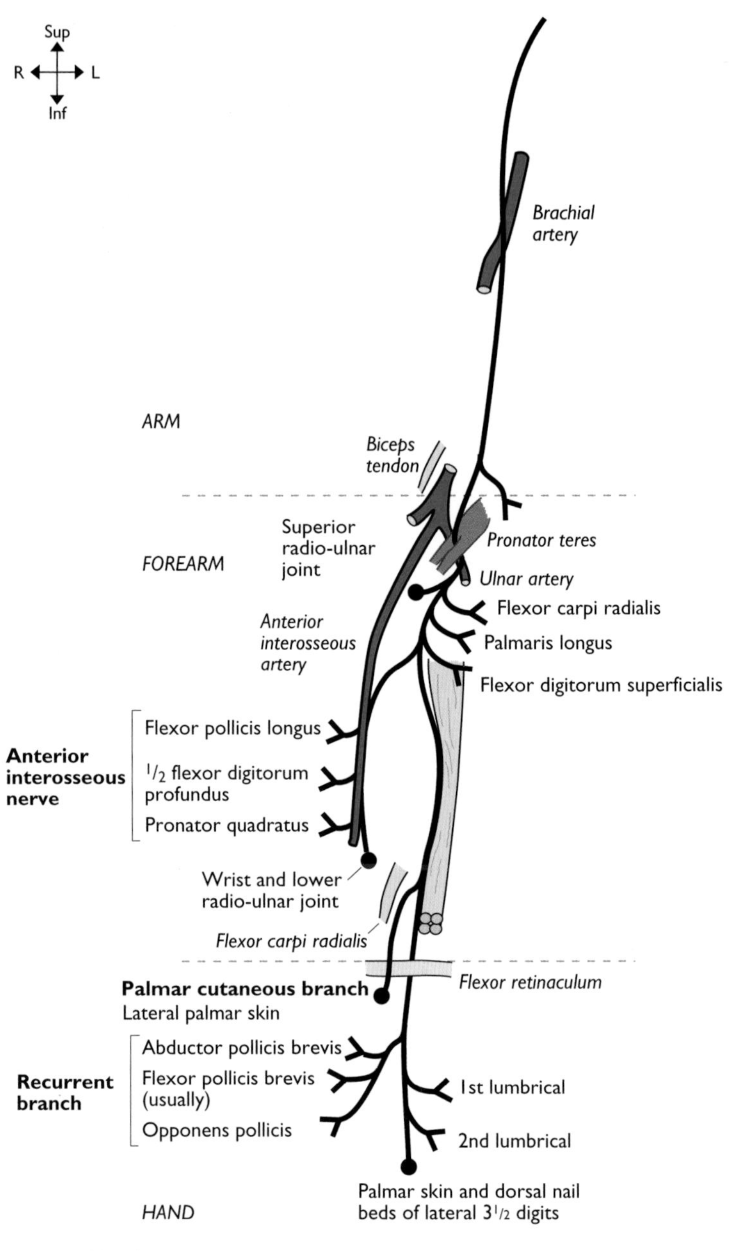

Median nerve (C6,7,8,T1)

MEDIAN NERVE (C6,7,8,T1)
From: **Med & lat cords**
To: **Terminal brs**

It is formed in the lower axilla by two 'heads' (contributions from medial and lateral cords) which clasp the axillary artery. The nerve initially lies anterior to the axillary artery and then lateral to it and subsequently lateral to the brachial artery. The median N then crosses the brachial artery, usually anteriorly, at the level of the mid humerus, to lie medial to the artery in the cubital fossa. It lies first on coracobrachialis and then brachialis. It passes beneath the bicipital aponeurosis at the elbow leaving the cubital fossa between the two heads of pronator teres before crossing superficial to the ulnar artery and giving its anterior interosseous branch below this. It lies applied to the deep surface of flexor digitorum superficialis on flexor digitorum profundus. It emerges from the lateral side of flexor digitorum superficialis about 5 cm proximal to the wrist where it gives its palmar cutaneous branch and then passes deep to the flexor retinaculum between the tendons of flexor digitorum superficialis and flexor carpi radialis. In the carpal tunnel it divides into terminal branches—recurrent (muscular) branch and palmar digital Ns.

Anterior interosseous N. Arises just below the two heads of pronator teres to run on the interosseous membrane between and covered by flexor digitorum profundus and flexor pollicis longus ending beneath pronator quadratus.

Recurrent branch of median N. Runs out of the carpal tunnel over the distal border of the flexor retinaculum onto flexor pollicis brevis to terminate by passing into the thenar eminence.

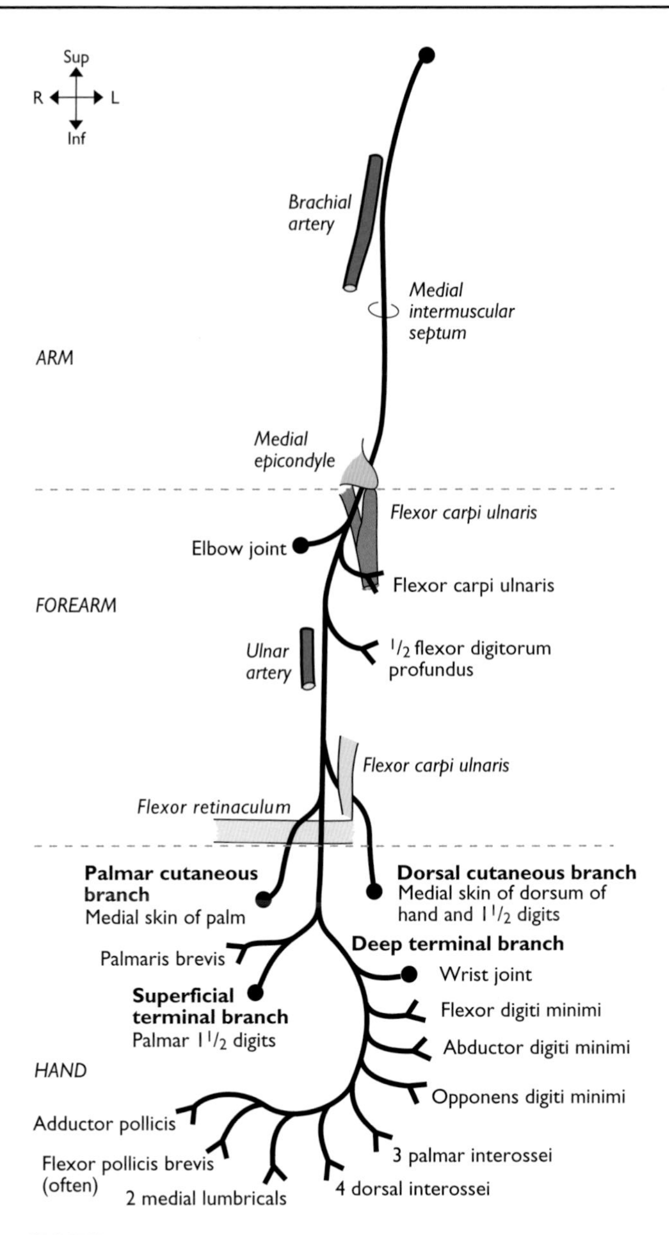

Ulnar nerve (C8,T1)

ULNAR NERVE (C8,T1)
From: **Med cord of brachial plexus**
To: **Terminal brs**

It arises medial to the axillary artery and continues medial to the brachial artery lying on coracobrachialis to the mid point of the humerus where it leaves the anterior compartment by passing posteriorly through the medial intermuscular septum with the superior ulnar collateral artery. It lies between the intermuscular septum and the medial head of triceps passing posterior to the medial humeral epicondyle and enters the forearm between the two heads of flexor carpi ulnaris. It then lies medial to the coronoid process of the ulna, runs deep to flexor carpi ulnaris and on flexor digitorum profundus, with the ulnar artery on its lateral side from one-third of the way down the forearm. It lies lateral to the tendon of flexor carpi ulnaris at the wrist and then passes superficial to the flexor retinaculum to divide into terminal branches at the pisiform bone.

Superficial terminal branch—lies superficial in the palm terminating as digital Ns. It also supplies palmaris brevis if present.

Deep terminal branch—passes through the hypothenar eminence between flexor digiti minimi brevis and abductor digiti minimi grooving the hook of the hamate and runs with the deep palmar arch, deep to the flexor tendons to terminate in adductor pollicis.

Dorsal cutaneous branch—arises 5 cm proximal to the wrist, passes deep to flexor carpi ulnaris onto the medial aspect of the dorsum of the hand where it terminates as cutaneous Ns.

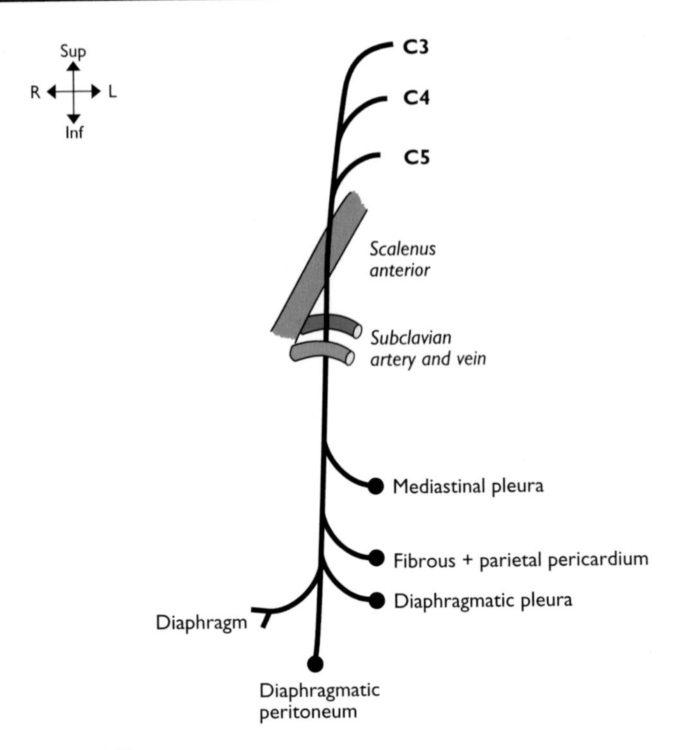

Phrenic nerve (C3,4,5)

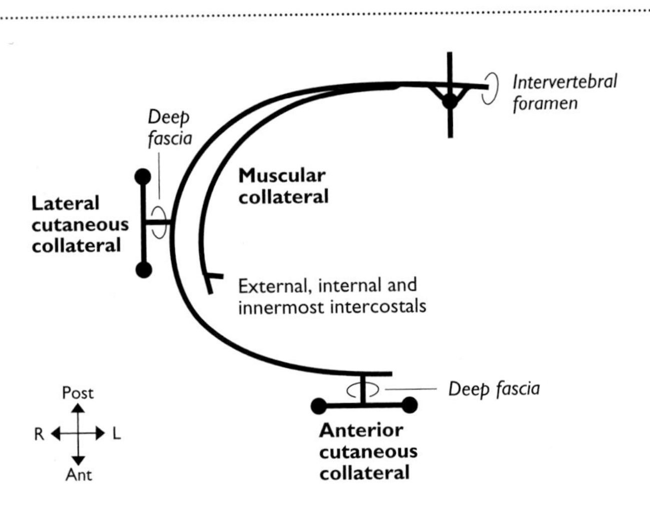

Intercostal nerve

PHRENIC NERVE (C3,4,5)
From: **Ant primary rami of C3,4,5**
To: **Terminal brs**

It arises deep between scalenus medius and scalenus anterior and runs over the lateral border of scalenus anterior behind the prevertebral fascia. It runs on scalenus anterior from lateral to medial edges lying lateral to the ascending cervical artery and it passes behind the suprascapular and transverse cervical arteries as it does so. It runs over the anterior part of the dome of the pleura to enter the mediastinum posterior to the subclavian vein and anterior to subclavian artery where right and left nerves take different courses. Right—spirals forwards to lie lateral to the right brachiocephalic vein and continues on the lateral surface of the superior vena cava, right atrium and inferior vena cava, lying within the fibrous pericardium. It passes anterior to the hilum of the lung before traversing the diaphragm via the caval orifice. Left—descends usually anterior to the left internal thoracic artery lying lateral to the left common carotid artery. It runs down over the aortic arch crossing anterior to the left vagus before running anterior to the left pulmonary artery and thus anterior to the hilum of the lung. It then runs lateral to the left auricle and left ventricle within the fibrous pericardium to traverse the diaphragm in isolation via the muscular portion of the diaphragm to the left of the central tendon.

An accessory phrenic N (C5) arising from the N to subclavius may join the phrenic N near the 1st rib.

INTERCOSTAL NERVE (TYPICAL)
From: **Ant primary rami of thoracic N**
To: **Terminal brs**

It emerges from the intervertebral foramen (giving off the posterior primary ramus as it does so) to pass between the pleura and the inner muscle layer anterior to the transverse process where it connects via the grey and white rami communicantes with the thoracic sympathetic chain. It passes posterior to the intercostal artery to lie below it as it runs in the subcostal groove in the plane between internal and innermost muscle layers. The muscular collateral branch arises before the posterior angle of the rib and runs in the same plane but lies at the level of the upper border of the rib below.

T1. Has no lateral or anterior cutaneous branches.

T7–11. Run behind the costal margins in their anterior course to lie in the same muscle plane in the abdomen. At their anterior limit they pass deep to rectus abdominis in the rectus sheath and pierce both of these structures to give terminal anterior cutaneous branches. These also supply rectus abdominis.

T12. Passes below the 12th rib as the subcostal N having similar branches to those above. It emerges from behind the lateral arcuate ligament of the diaphragm.

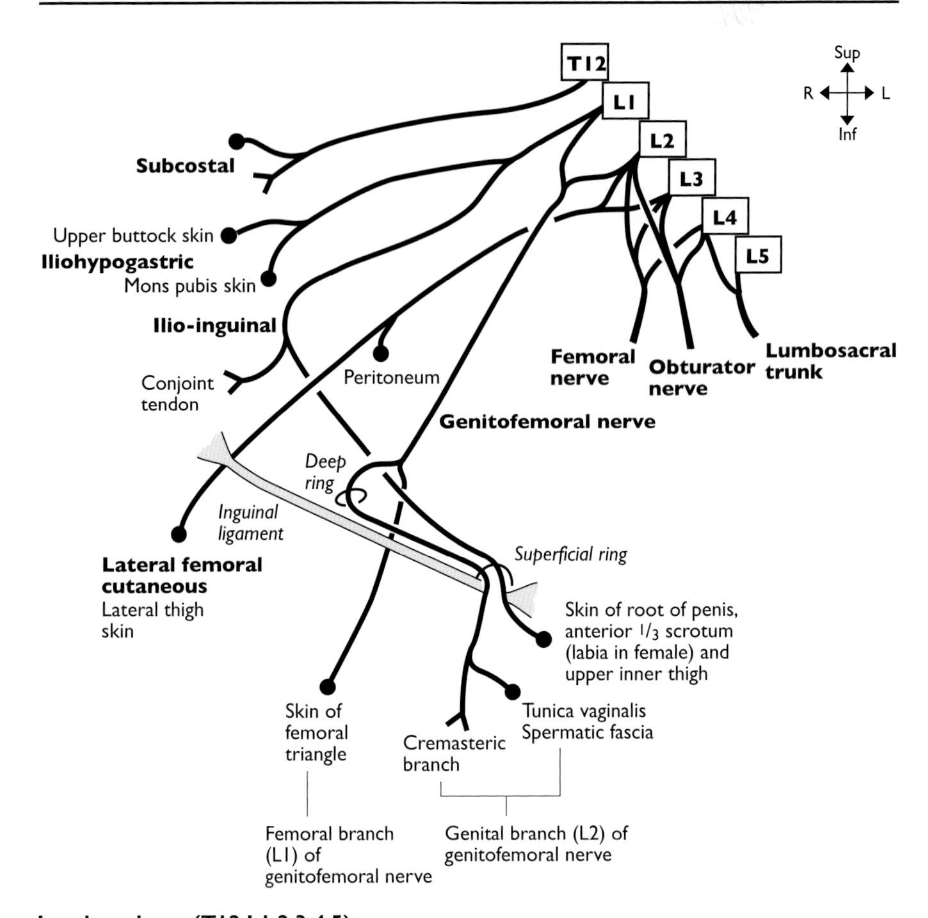

Lumbar plexus (T12,L1,2,3,4,5)

LUMBAR PLEXUS (T12,L1,2,3,4,5)
From: Ant primary rami of T12,L1,2,3,4,5
To: Ns as shown

The plexus is formed within the substance of psoas major anterior to the transverse processes of L2–5 from the anterior primary rami as they emerge from the intervertebral foramina.

Iliohypogastric N (L1). This main nerve emerges lateral to psoas on the lumbar fascia at the level of L2 to pass posterior to the lower pole of the kidney and over quadratus lumborum. It passes above the iliac crest, between transversus and internal oblique abdominis to pierce the latter above the anterior superior iliac spine. It supplies both muscles before becoming cutaneous.

Ilio-inguinal N (L1). This collateral branch emerges lateral to psoas on the lumbar fascia, passes posterior to the lower pole of the kidney, over quadratus lumborum and penetrates transversus and internal oblique abdominis above the anterior superior iliac spine. It supplies the lowest fibres of these muscles and the conjoint tendon. Its terminal branch enters the inguinal canal from above to pass through the superficial inguinal ring before piercing the external spermatic fascia to become subcutaneous.

Lateral femoral cutaneous N (L2,3). Emerges lateral to psoas below the iliac crest, passes over iliacus obliquely lying posterior to the caecum on the right and descending colon on the left. It runs forward to the anterior superior iliac spine where it penetrates the inguinal ligament at its attachment to pass into the subcutaneous tissue of the lateral thigh.

Genitofemoral N (L1,2). Emerges onto the anteromedial surface of psoas lying posterior to the ureter, gonadal and iliocolic vessels on the right and ureter, gonadal and lower left colic vessels on the left. It divides into genital and femoral branches on the anterior aspect of psoas.

Genital branch. Crosses the external iliac artery, passes through the deep inguinal ring into the inguinal canal and through the superficial inguinal ring to terminate in the spermatic cord in the male and labium majus in the female. It may also supply a small area of scrotal/labial skin.

Femoral branch. Continues down lateral to the external iliac artery, under the inguinal ligament and into the femoral sheath which it penetrates anteriorly to become subcutaneous.

Lumbosacral trunk (L4,5). Emerges deep from the medial aspect of psoas to pass over the pelvic brim to form the upper fibres of the sciatic N.

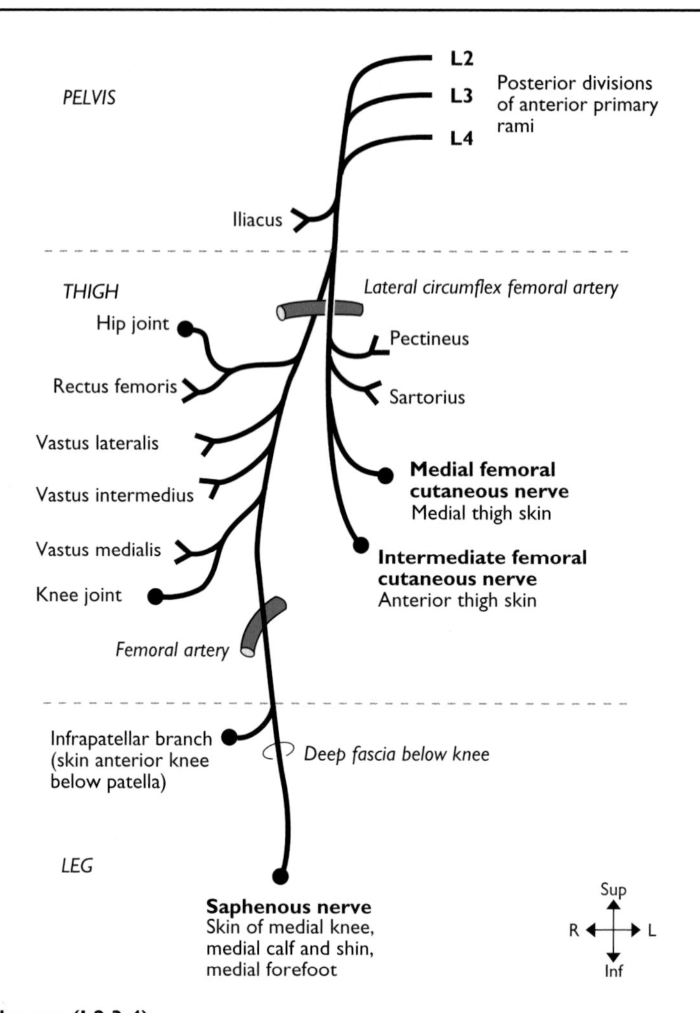

Femoral nerve (L2,3,4)

FEMORAL NERVE (L2,3,4)
From: **Post div of ant primary rami of L2,3,4**
To: **Terminal brs**

It is formed within psoas major and emerges from its lateral border low down in the iliac fossa to lie in the groove between psoas and iliacus. It reaches the thigh beneath the inguinal ligament lateral to the femoral artery lying on the tendon of iliacus and psoas. In the femoral triangle it splits into anterior and posterior divisions which straddle the lateral circumflex femoral artery. There are usually four short superficial branches. The deep branches continue down the femoral triangle, the N to vastus medialis running lateral to the femoral artery as far as the upper part of the adductor (Hunter's) canal before entering the muscle.

Saphenous N (post division). Descends in the femoral triangle to reach the adductor canal where it spirals over the femoral artery to lie medial to it. It pierces the deep fascia through the apex of the canal and emerges posterior to sartorius and anterior to gracilis to continue with the long saphenous vein. It passes over the subcutaneous surface of the tibia and anterior to the medial malleolus where it is palpable, closely related to the long saphenous vein. It terminates in branches over the medial side of the foot.

6

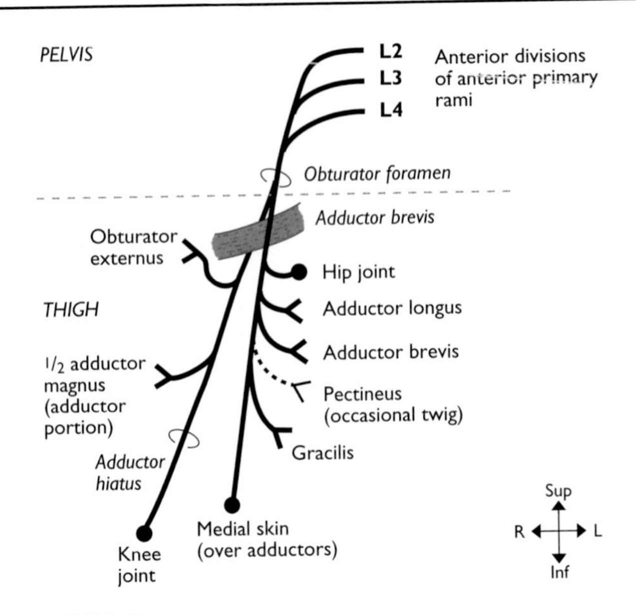

6

Obturator nerve (L2,3,4)

OBTURATOR NERVE (L2,3,4)
From: **Ant div of ant primary rami of L2,3,4**
To: **Terminal brs**

This nerve is formed within psoas major and emerges from the medial aspect of the muscle on the ala of the sacrum to pass behind the common iliac vessels. It runs over the pelvic brim on the lateral wall of the pelvis and over the upper fibres of obturator internus to pass through the upper anterior aspect of the obturator foramen. It divides into anterior and posterior divisions which straddle adductor brevis. The posterior division pierces a few fibres of obturator externus and runs deep to adductor brevis on adductor magnus. The anterior division runs on the anterior aspect of adductor brevis deep to pectineus and then deep to adductor longus to end by contributing, together with the saphenous and medial femoral cutaneous N, to the subsartorial plexus which supplies the skin over the medial thigh.

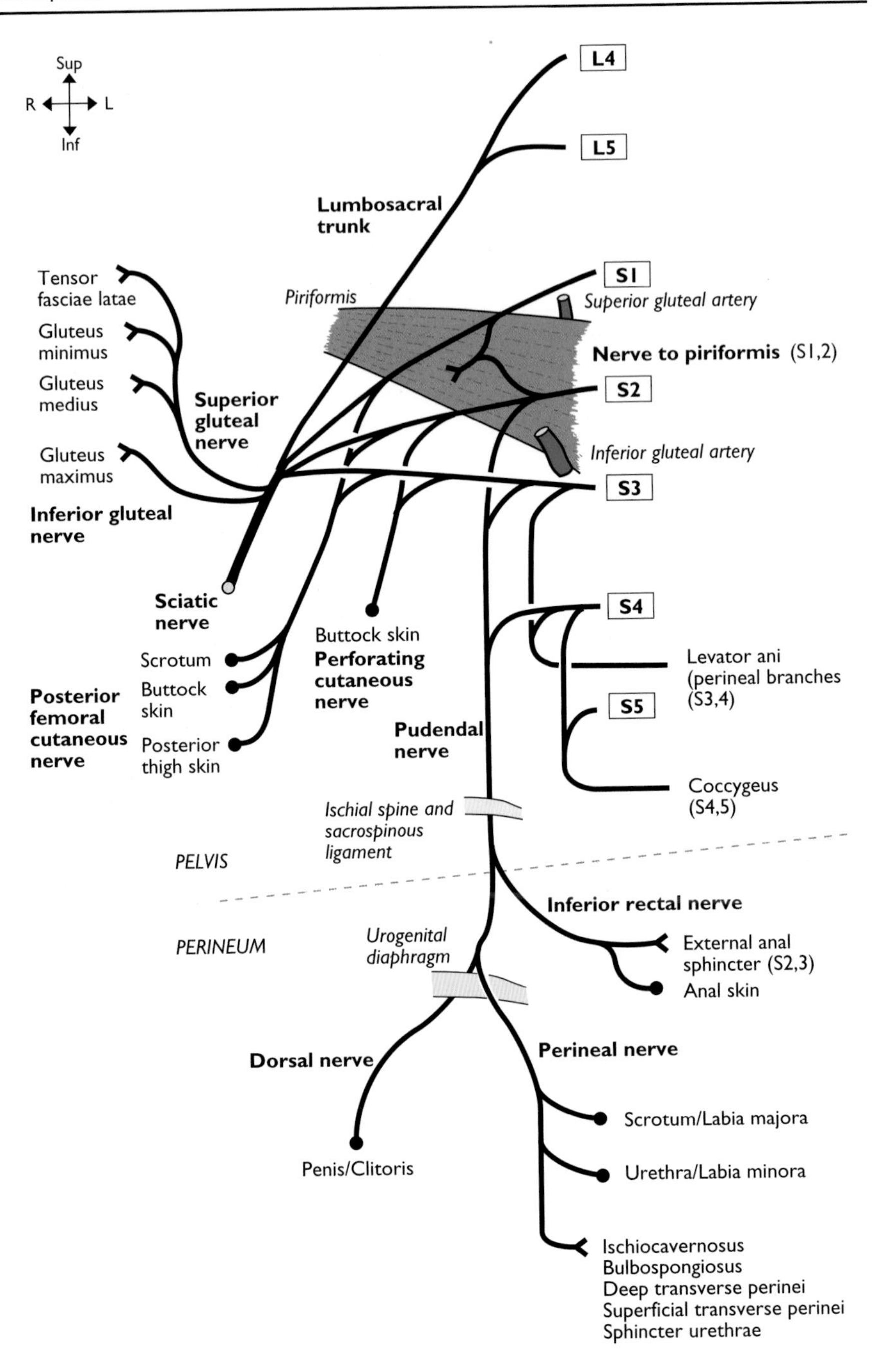

Sacral plexus (L4,5,S1,2,3,4,5)

SACRAL PLEXUS (L4,5,S1,2,3,4,5)
From: **Lumbosacral trunk (L4,5) & ant primary rami from S1,2,3,4,5**
To: **Definitive Ns**

Lies on piriformis on the posterior wall of the pelvis deep to the internal iliac vessels (and the sigmoid vessels on the left) and is protected by a sheet of pelvic fascia overlying it. Its roots are characteristically related to arteries which pass between them as shown.

Superior gluteal N (L4,5,S1). Emerges from the upper roots of the sciatic N and passes out of the pelvis above piriformis through the greater sciatic foramen. It runs between gluteus medius and minimus over the middle gluteal line on the outer surface of the ilium to terminate in muscular branches.

Inferior gluteal N (L5,S1,2). Emerges from the middle roots of the sciatic N and passes out of the pelvis below piriformis through the greater sciatic foramen to enter gluteus maximus.

Posterior femoral cutaneous N (S1,2,3). Passes out of the pelvis below piriformis through the greater sciatic foramen. It runs on the sciatic N, over the long head of biceps femoris to become subcutaneous extending as far as the popliteal fossa.

Perforating cutaneous N (S2,3). Passes through the sacrotuberous ligament and inferior to gluteus maximus to become subcutaneous in the buttock. It supplies that part of the buttocks which touch when standing and part when sitting.

Pudendal N (S2,3,4). Passes out of the pelvis over the sacrospinous ligament close to the ischial spine through the greater, and re-entering through the lesser, sciatic foramina. It runs on the medial surface of the lower fibres of obturator internus in the pudendal (Alcock's) canal. It passes forwards in the lateral wall of the ischio-anal fossa where it gives off its inferior rectal branch. It passes into the perineum and gives its terminal branches, the perineal N being superficial to the urogenital diaphragm and the dorsal N deep to it.

It helps to remember that all six branches leaving sacral roots before anterior and posterior divisions begin with the letter 'P':
N to Piriformis S1,2 (remains in pelvis to supply this muscle)
Posterior femoral cutaneous N S1,2,3 (leaves pelvis via greater sciatic foramen)
Perforating cutaneous N S2,3 (leaves pelvis via greater sciatic foramen)
Pudendal N S2,3,4 (leaves pelvis via greater sciatic foramen)
Pelvic splanchnics (parasympathetic) S2,3,4 (remain in pelvis to supply pelvic organs)
Perineal branch of S4 (remains in pelvis to supply levator ani)

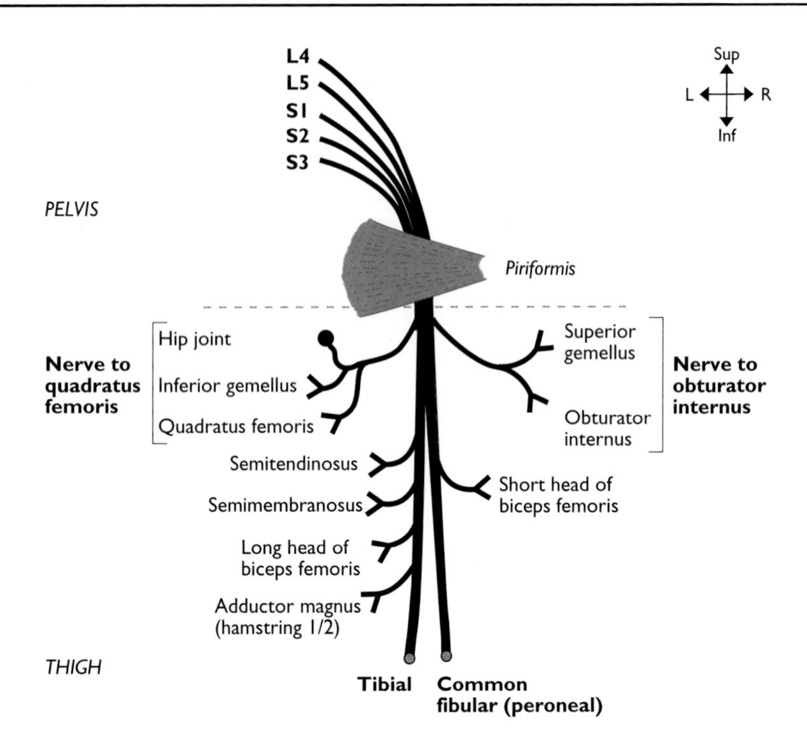

Viewed from behind

Sciatic nerve (L4,5,S1,2,3)

SCIATIC NERVE (L4,5,S1,2,3)

From: **Ant primary rami of L4,5,S1,2,3**
To: **Tibial & common fibular (peroneal) Ns**

It is formed in the upper sacral plexus and passes out of the greater sciatic foramen below piriformis. In the buttock and thigh it lies initially deep to gluteus maximus lying on gemellus superior, obturator internus tendon and gemellus inferior and then on quadratus femoris and adductor magnus. It passes out of the cover of gluteus maximus and for a short distance it is covered by only deep fascia, before it passes deep to the two heads of biceps femoris. It runs vertically down in the midline of the posterior compartment of the thigh and terminates by dividing into common fibular (peroneal) and tibial Ns usually two-thirds of the way down the thigh. In its course over the gemelli it is a close posterior relation of the ischium and posterior rim of the acetabulum.

N to quadratus femoris (L4,5,S1). Arises from the anterior surface of the sciatic N in the pelvis and leaves the pelvis in this position through the greater sciatic foramen, lying between the sciatic N and the ischium. Running deep to the tendon of obturator internus and the gemelli it supplies gemellus inferior before passing into quadratus femoris from above.

N to obturator internus (L5,S1,2). Arises from the anterior surface of the sciatic N in the pelvis and leaves the pelvis through the greater sciatic foramen below piriformis and medial to the sciatic N. It passes medially over the ischial spine (lateral to the pudendal neurovascular bundle) and sends a branch to gemellus superior before turning forward to pass through the lesser sciatic foramen, penetrating and supplying obturator internus as it does so.

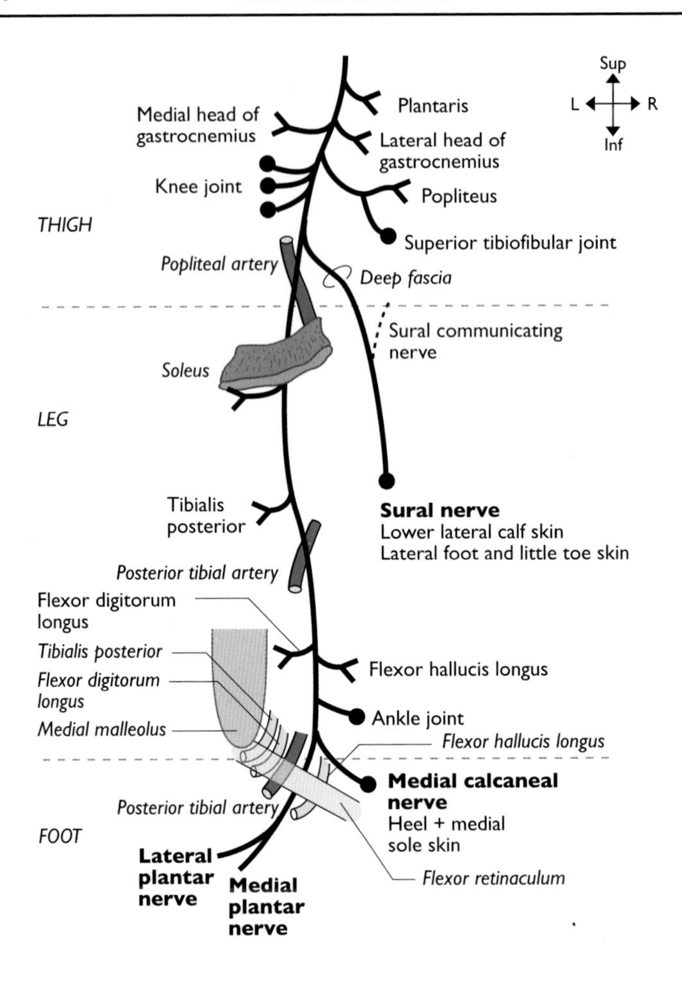

Tibial nerve (L4,5,S1,2,3)

TIBIAL NERVE (L4,5,S1,2,3)
From: **Sciatic N**
To: **Med & lat plantar Ns**

It arises in the lower third of the thigh above the apex of the popliteal fossa as the larger terminal branch of the sciatic N, and passes down in the midline into the fossa between semitendinosus and biceps femoris, lying deep to them. It lies markedly lateral to the popliteal artery on entry to the popliteal fossa but then the artery crosses deep to the N to lie lateral to it. The tibial N and the popliteal artery remain separated by the popliteal veins. The nerve leaves the fossa deep to the two heads of gastrocnemius by passing deep to the fibrous arch of soleus. It then runs deep to soleus on tibialis posterior in the midline, crossing over the posterior tibial artery from medial to lateral half way down the calf. It slopes gently medially in the lower calf passing behind the medial malleolus of the lower tibia between the posterior tibial artery anteromedially and the tendon of flexor hallucis longus posterolaterally. It runs under the flexor retinaculum where it divides into terminal branches.

Sural N. Arises in the popliteal fossa, passing out posteriorly over the 'V' behind the two heads of gastrocnemius and is joined by the sural communicating N from the common fibular (peroneal) N. It pierces the deep fascia to become subcutaneous. It runs down laterally accompanied by the short saphenous vein to pass behind the lateral malleolus over the superior fibular (peroneal) retinaculum to end in terminal branches on the lateral side of the foot.

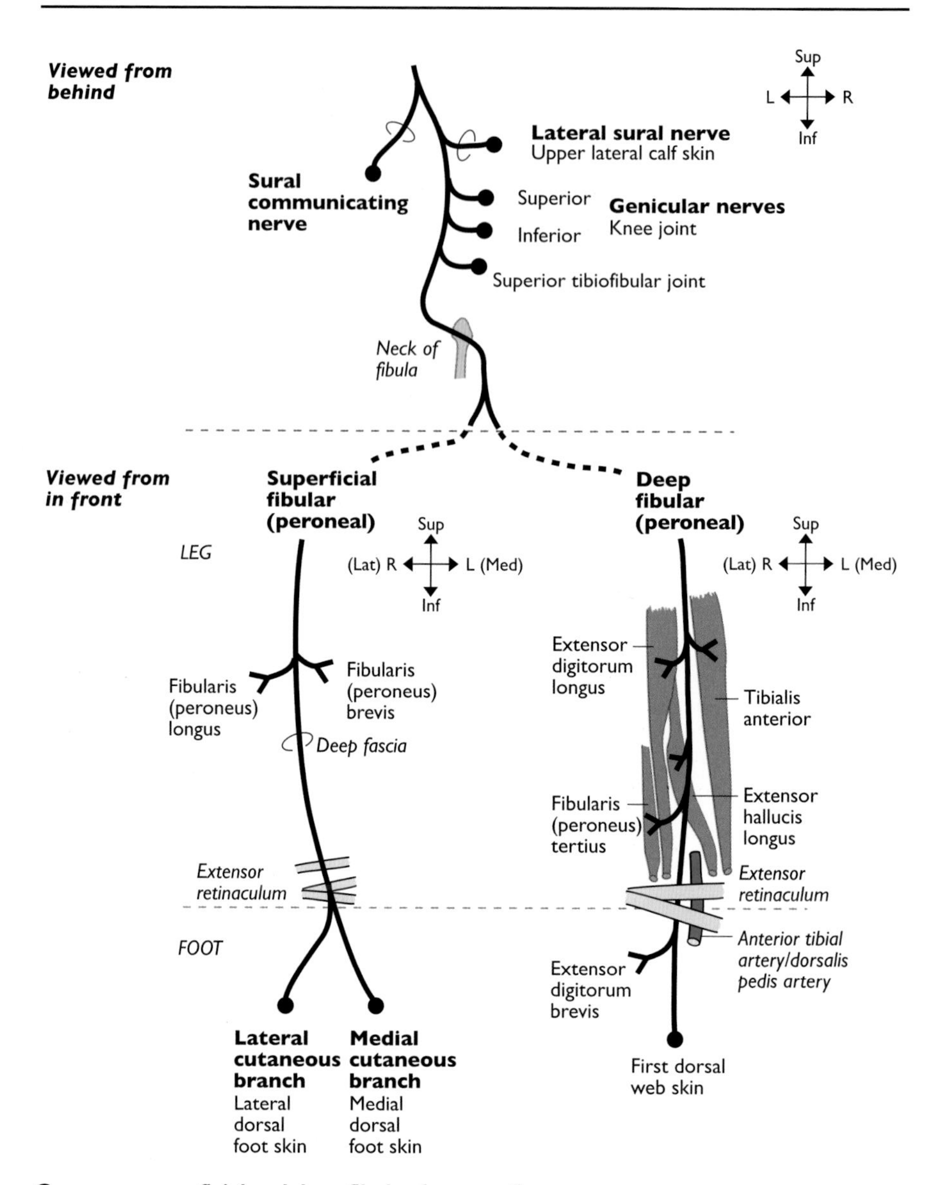

Common, superficial and deep fibular (peroneal) nerves
Note: the common fibular (peroneal) nerve is viewed from behind but the superficial and deep nerves are viewed from in front

COMMON FIBULAR [PERONEAL] NERVE (L4,5,S1,2)
From: Sciatic N
To: Superficial & deep fibular (peroneal) Ns

It arises in the lower third of the thigh above the apex of the popliteal fossa as the smaller terminal branch of the sciatic N. It passes into the popliteal fossa along the upper lateral boundary just beneath the edge of biceps femoris and runs over plantaris, the lateral head of gastrocnemius and the posterior capsule of the knee joint. It runs over the fibular attachment of soleus to wind around the neck of the fibula from posterior to lateral. It passes into fibularis (peroneus) longus where it divides.

SUPERFICIAL FIBULAR [PERONEAL] NERVE (L5,S1,2)
From: Common fibular (peroneal) N
To: Terminal brs

It arises deep to fibularis (peroneus) and passes forwards and downwards to lie over the lateral surface of the fibula between fibularis (peroneus) longus and brevis. It pierces the deep fascia half way down the leg to become subcutaneous. It runs downwards superficial to the superior and inferior extensor retinacula to end as terminal branches over them.

DEEP FIBULAR [PERONEAL] NERVE (L4,5,S1,2)
From: Common fibular (peroneal) N
To: Terminal brs

It arises deep to fibularis (peroneus) longus and passes forwards deep to the muscle to wind around the fibula and to pass through the anterior intermuscular septum. It continues deep to extensor digitorum longus to appear between it and tibialis anterior lying on the interosseous membrane in the upper quarter of the anterior compartment. It runs down the interosseous membrane with the anterior tibial vessels, coming to lie between extensor hallucis longus and tibialis anterior in the lower three-quarters of the compartment. It passes anterior to the tibia at the ankle joint between the anterior tibial artery medially and the tendon of extensor digitorum longus laterally, running beneath the superior and inferior extensor retinacula. It breaks up into terminal branches on the dorsum of the foot. (Articular branch to ankle joint not shown.)

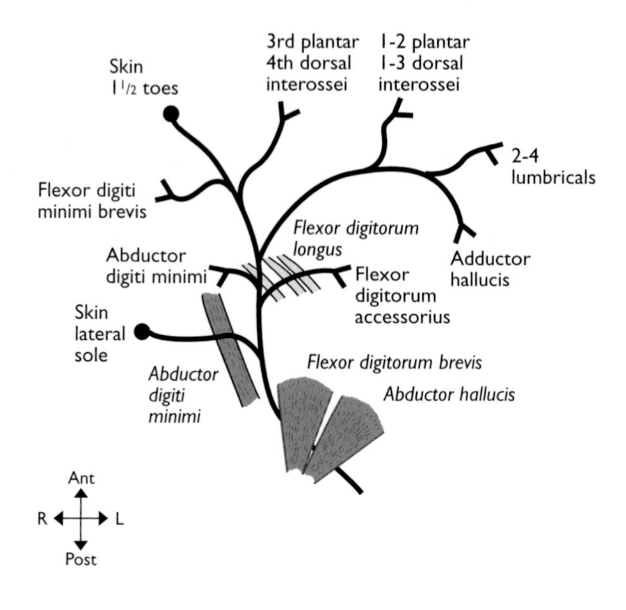

6

Lateral plantar nerve (S1, 2, 3)
Skin S1
All muscles S2, 3

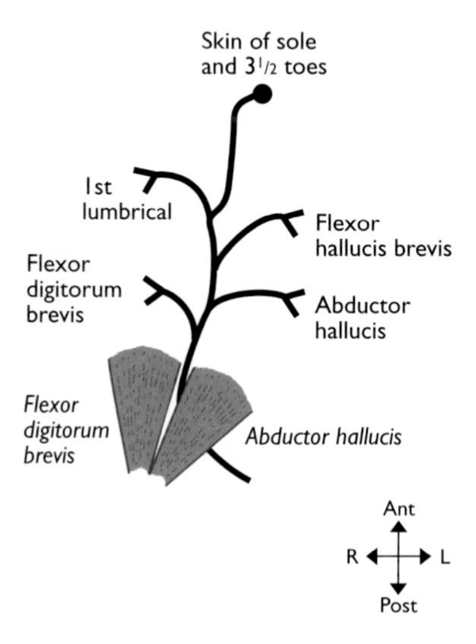

Medial plantar nerve (L4, 5, S1, 2, 3)
Skin L4, 5
All muscles S1, 2 except
1st lumbrical which is S2, 3

LATERAL PLANTAR NERVE (S1,2,3)
From: Tibial N
To: Terminal brs

It arises beneath the flexor retinaculum and runs forward with the lateral plantar artery around the sustentaculum tali of the calcaneus deep to abductor hallucis. It runs over the origin of flexor digitorum accessorius beneath flexor digitorum brevis, and its superficial terminal branches appear more superficially between flexor digitorum brevis and abductor digiti minimi. Its deep terminal branches run medially beneath the long flexor tendons and across the metatarsal shafts to end in muscular branches.

6

MEDIAL PLANTAR NERVE (L4,5,S1,2,3)
From: Tibial N
To: Terminal brs

It arises beneath the flexor retinaculum and runs with the medial plantar artery around the sustentaculum tali of the calcaneus deep to abductor hallucis. It pierces the plantar fascia in so doing and runs forward over the tendon of flexor digitorum longus to appear more superficially again between abductor hallucis and flexor digitorum brevis in the sole of the foot.

7: DERMATOMES AND CUTANEOUS NERVE DISTRIBUTION

7

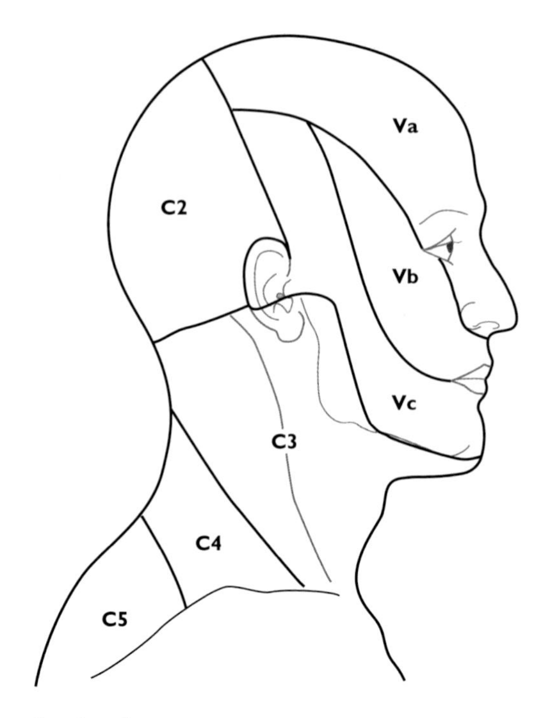

Dermatomes: head and neck

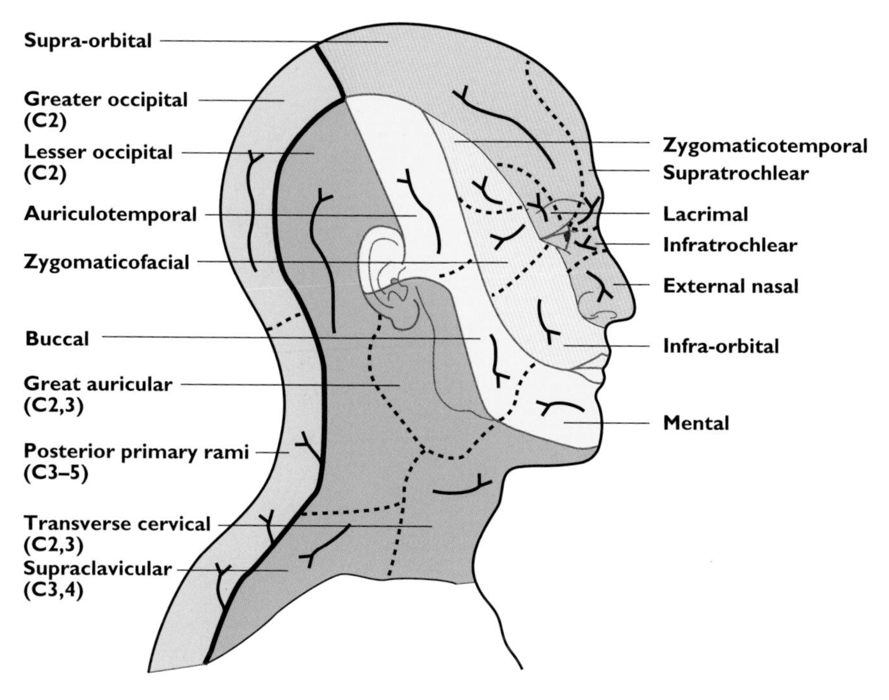

Cutaneous nerves: head and neck

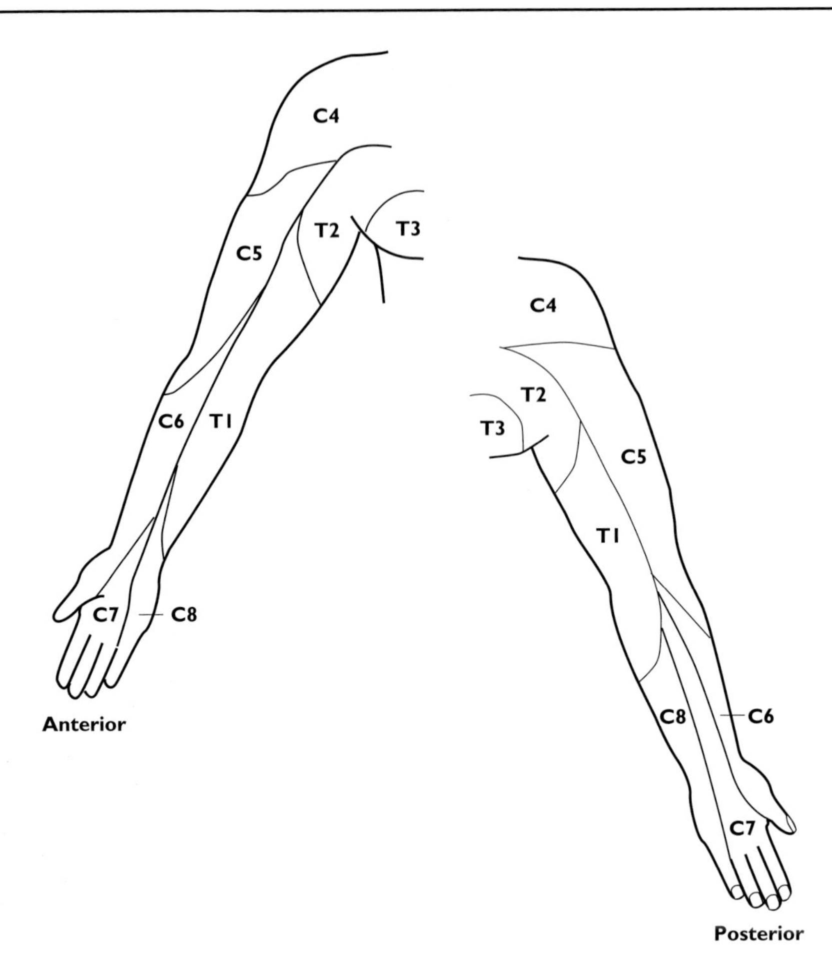

Dermatomes: upper limb

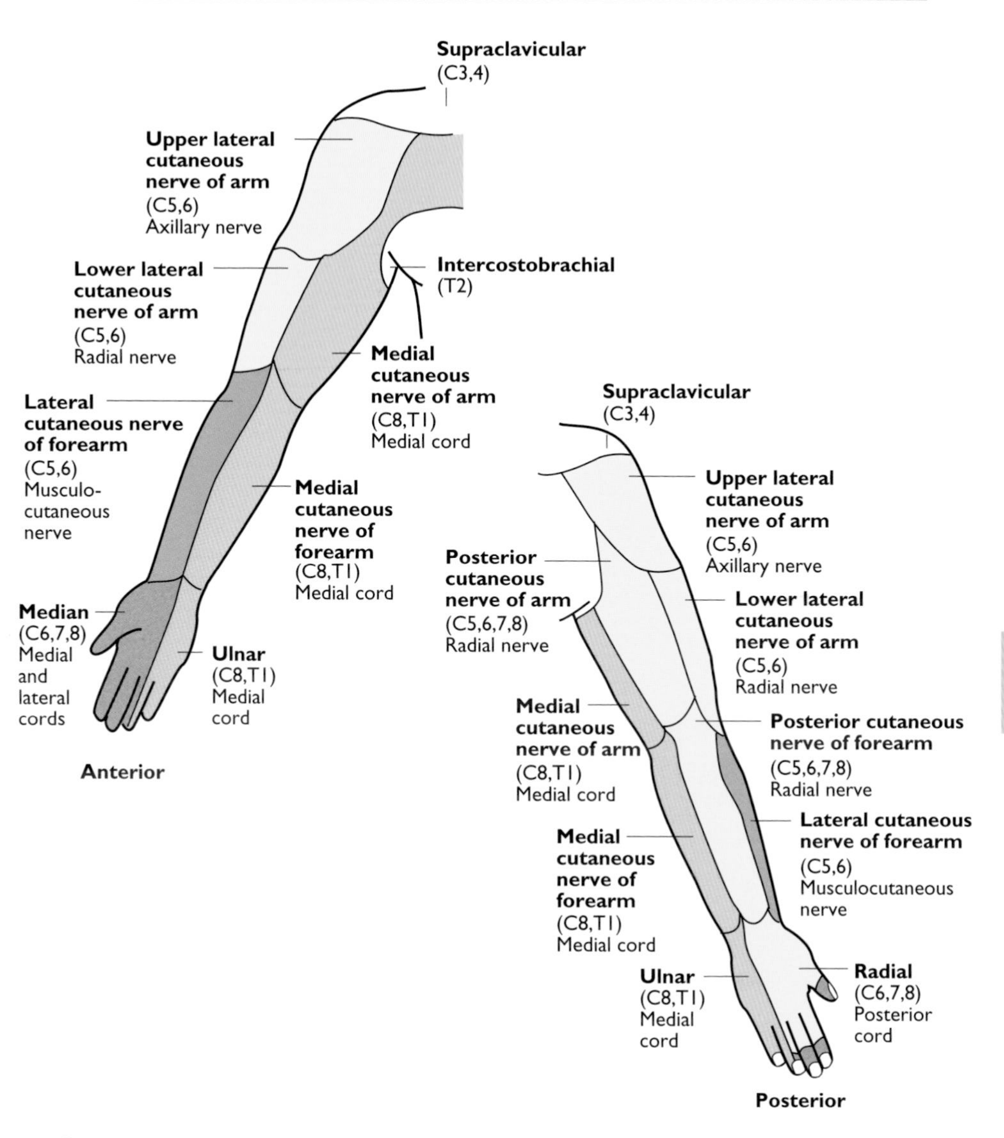

Cutaneous nerves: upper limb

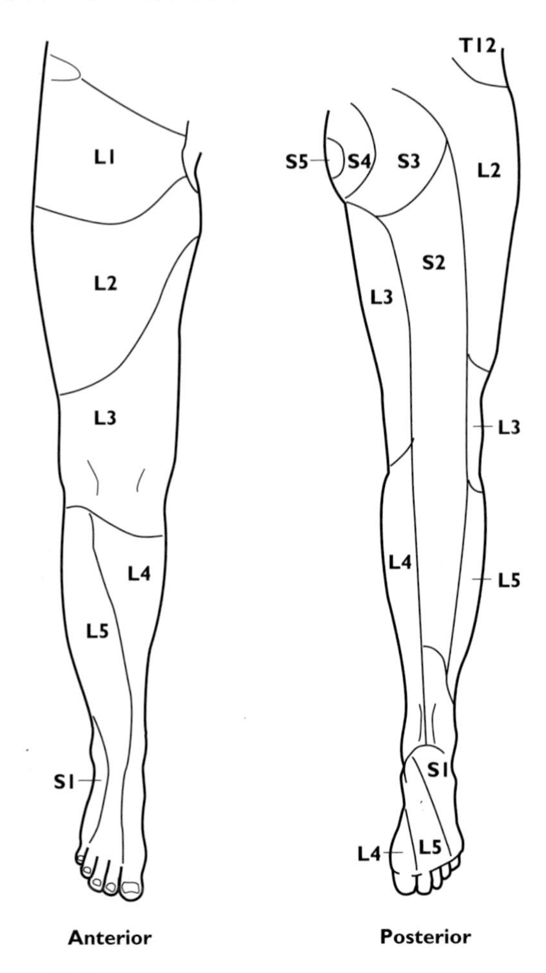

Dermatomes: lower limb

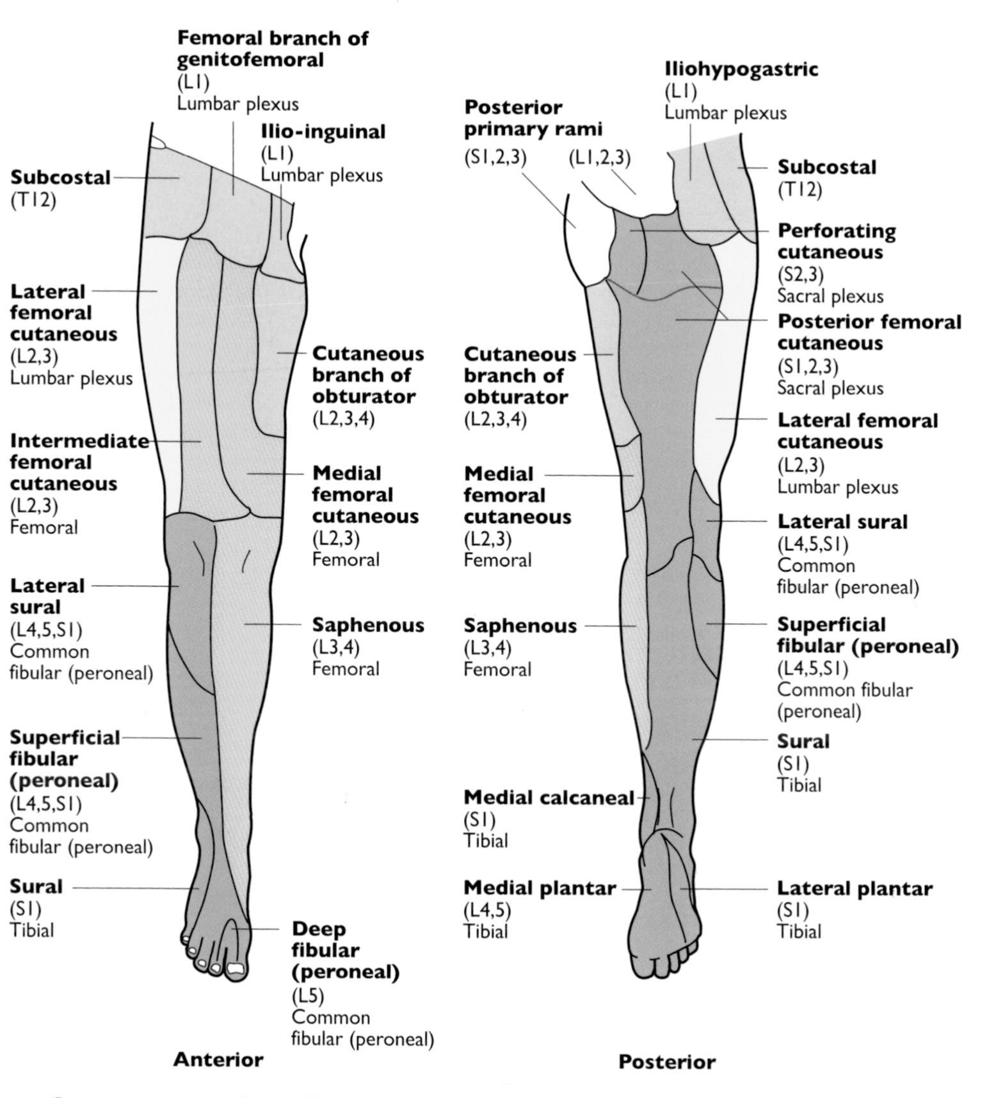

Cutaneous nerves: lower limb

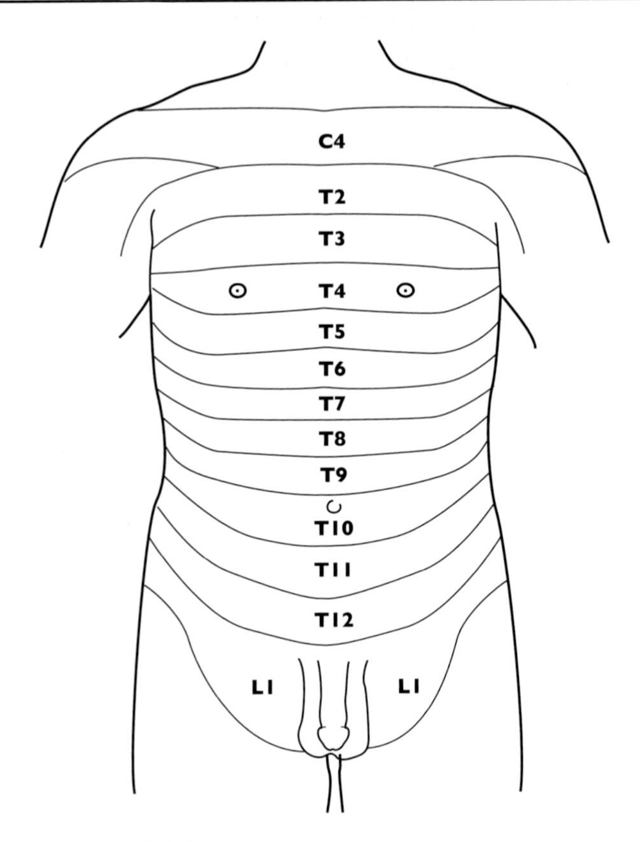

Dermatomes: thorax and abdomen

8: MUSCLES

ABDUCTOR DIGITI MINIMI (foot)
Arises Med & lat processes of post calcaneal tuberosity
Inserts Lat side of base of prox phalanx of 5th toe & 5th MT
Action Flexes & abducts 5th toe. Supports lat longitudinal arch
Nerve Lat plantar N (S2,3)

ABDUCTOR DIGITI MINIMI (hand)
Arises Pisiform bone & pisohamate lig
Inserts Ulnar side of base of prox phalanx of little finger & extensor expansion (± sesamoid)
Action Abducts little finger at MCP jnt
Nerve Deep br of ulnar N (C8,T1)

ABDUCTOR HALLUCIS
Arises Med process of post calcaneal tuberosity & flexor retinaculum
Inserts Med aspect of base of prox phalanx of big toe via med sesamoid
Action Flexes & abducts big toe. Supports med longitudinal arch
Nerve Med plantar N (S1,2)

ABDUCTOR POLLICIS BREVIS
Arises Tubercle of scaphoid & flexor retinaculum
Inserts Radial sesamoid of prox phalanx of thumb & tendon of extensor pollicis longus
Action Abducts thumb at MCP & CMC jnts
Nerve Recurrent (muscular) br of median N (C8,T1)
Notes As this muscle is supplied only by median N, its action is used as a test for this N in hand

ABDUCTOR POLLICIS LONGUS
Arises Upper post surface of ulna & middle third of post surface of radius & interosseous membrane between
Inserts Over tendons of radial extensors & brachioradialis to base of 1st MC & trapezium
Action Abducts & extends thumb at CMC jnt
Nerve Post interosseous N (C7,8)
Notes Forms radial side of snuff box

ADDUCTOR BREVIS
Arises Inf ramus & body of pubis
Inserts Upper third of linea aspera
Action Adducts hip
Nerve Ant div of obturator N (L2,3)

ADDUCTOR HALLUCIS
Arises Oblique head: base of 2, 3, 4 MTs. Transverse head: plantar MT ligs & deep transverse lig
Inserts Lat side of base of prox phalanx of big toe & lat sesamoid
Action Adducts & flexes MTP jnt of big toe. Supports transverse arch
Nerve Deep br of lat plantar N (S2,3)
Notes If muscle fibres are attached to 1st MT it can be regarded as opponens hallucis

ADDUCTOR LONGUS
Arises Body of pubis inf & med to pubic tubercle
Inserts Lower two-thirds of med linea aspera
Action Adducts hip
Nerve Ant div of obturator N (L2,3,4)

ADDUCTOR MAGNUS
Arises Adductor portion: ischiopubic ramus. Hamstring portion: lower outer quadrant of post surface of ischial tuberosity
Inserts Adductor portion: lower gluteal line & linea aspera. Hamstring portion: adductor tubercle
Action Adductor portion: adducts hip. Hamstring portion: extends hip
Nerve Adductor portion: post div of obturator N (L2,3,4). Hamstring portion: tibial portion of sciatic (L4)

ADDUCTOR POLLICIS
Arises Oblique head: base of 2nd & 3rd MCs, trapezoid & capitate. Transverse head: palmar border & shaft of 3rd MC
Inserts Ulnar sesamoid then ulnar side of base of prox phalanx & tendon of extensor pollicis longus
Action Adducts CMC jnt of thumb
Nerve Deep br of ulnar N (C8,T1)

ANCONEUS
Arises Smooth surface at lower extremity of post aspect of lat epicondyle of humerus
Inserts Lat side of olecranon
Action Weak extensor of elbow. Moves (abducts) ulna in pronation
Nerve Radial N (C7,8)

ARTICULARIS CUBITI (subanconeus)
Arises Deep distal surface of med head of triceps
Inserts Post capsule of elbow jnt
Action Lifts capsule away from jnt
Nerve Radial (C6,7,8)

ARTICULARIS GENU
Arises Two slips from ant femur below vastus intermedius
Inserts Apex of suprapatellar bursa
Action Retracts bursa as knee extends
Nerve Post div of femoral N (L2,3,4)

ARYEPIGLOTTICUS
Arises Apex of arytenoid cartilage
Inserts Lat border of epiglottis
Action Aids closure of additus of larynx
Nerve Recurrent laryngeal br of vagus N (X)
Note It is an extension of oblique arytenoid

AURICULARIS
Arises Cartilage of auricle
Inserts Galeal aponeurosis
Action Adjusts position of ear
Nerve Temporal & posterior auricular brs of facial N (VII)

BICEPS BRACHII
Arises Long head: supraglenoid tubercle of scapula. Short head: coracoid process of scapula with coracobrachialis
Inserts Post border of bicipital tuberosity of radius (over bursa) & bicipital aponeurosis to deep fascia & subcutaneous ulna
Action Supinates forearm, flexes elbow, weakly flexes shoulder
Nerve Musculocutaneous N (C5,6) (from lat cord)
Notes Supinates most effectively when elbow flexed

BICEPS FEMORIS
Arises Long head: upper inner quadrant of post surface of ischial tuberosity. Short head: middle third of linea aspera, lat supracondylar ridge of femur
Inserts Styloid process of head of fibula, lat collateral lig & lat tibial condyle
Action Flexes & lat rotates knee. Long head extends hip
Nerve Long head: tibial portion of sciatic N. Short head: common fibular (peroneal) portion of sciatic N (both L5,S1,2)

BRACHIALIS
Arises Ant lower half of humerus & med & lat intermuscular septa
Inserts Coronoid process & tuberosity of ulna
Action Flexes elbow
Nerve Musculocutaneous N (C5,6) (from lat cord). Also small supply from radial N (C7)

BRACHIORADIALIS
Arises Upper two-thirds of lat supracondylar ridge of humerus & lat intermuscular septum
Inserts Base of styloid process of radius
Action Flexes arm at elbow & brings forearm into midprone position
Nerve Radial N (C5,6)
Notes Overlies radial N & art as they lie on supinator

BUCCINATOR

Arises Ext alveolar margins of maxilla & mandible by molar teeth, to maxillary tubercle & pterygoid hamulus & post mylohyoid line respectively, then via pterygomandibular raphe between bones
Inserts Decussates at modiolus of mouth & interdigitates with opposite side
Action Aids mastication by emptying vestibule, tenses cheeks in blowing & whistling, aids closure of mouth
Nerve Buccal br of facial N (VII)

BULBOSPONGIOSUS

Arises Perineal body (& midline raphe over corpus spongiosum in male)
Inserts Superficial perineal membrane & dorsal penile/clitoral aponeurosis
Action Male: aids emptying of urine & ejaculate from urethra. Female: closes vaginal introitus
Nerve Perineal br of pudendal N (S2,3,4)

8

CONSTRICTOR OF PHARYNX—INFERIOR

Arises Cricopharyngeus: lat aspect of arch of cricoid cartilage. Thyropharyngeus: oblique line on laminar of thyroid cartilage & fibrous cricothyroid arch
Inserts Cricopharyngeus: continuous with muscle of opposite side, behind pharynx. Thyropharyngeus: lower pharyngeal raphe
Action Aids swallowing. Cricopharyngeus acts as upper oesophageal sphincter
Nerve Pharyngeal plexus (IX, X & sympathetic) via pharyngeal br of X with its motor fibres from cranial accessory (XI)
Notes Killian's dehiscence is between the two parts post

CONSTRICTOR OF PHARYNX—MIDDLE

Arises Lower third of stylohyoid lig, lesser cornu & sup border of greater cornu of hyoid bone
Inserts Middle portion of pharyngeal raphe
Action Aids swallowing
Nerve Pharyngeal plexus (IX, X & sympathetic) via pharyngeal br of X with its motor fibres from cranial accessory N (XI)

CONSTRICTOR OF PHARYNX—SUPERIOR

Arises Lower two-thirds of med pterygoid plate, pterygomandibular raphe & post end of mylohyoid line on mandible
Inserts Upper midline pharyngeal raphe & pharyngeal tubercle of occiput
Action Aids swallowing
Nerve Pharyngeal plexus (IX, X & sympathetic) via pharyngeal br of X with its motor fibres from cranial accessory (XI)

CORACOBRACHIALIS

Arises Coracoid process of scapula with biceps brachii
Inserts Middle third of med border of humerus
Action Flexes & weakly adducts arm
Nerve Musculocutaneous N (C5,6,7) (from lat cord)
Notes Ligament of Struthers as embryological 3rd head. Musculocutaneous N runs through muscle

CORRUGATOR SUPERCILII

Arises Med superciliary arch
Inserts Skin of med forehead
Action Wrinkles forehead
Nerve Temporal br of facial N (VII)

CREMASTER

Arises Lower border of internal oblique & transversus abdominis in inguinal canal
Inserts Loops around spermatic cord & tunica vaginalis & some fibres return to attach to pubic tubercle
Action Retracts testis
Nerve Sympathetic & somatic fibres in genital br (L2) of genitofemoral N (L1,2)

CRICOTHYROID
Arises Anterolateral aspect of cricoid cartilage
Inserts Inf cornu & lower laminar of thyroid cartilage
Action Lengthens & tenses vocal cords by tilting thyroid cartilage forwards
Nerve Ext br of sup laryngeal br of vagus N (X)

DARTOS
Arises Subcutaneous tissue of scrotum, superficial to superficial fascia (Colles)
Inserts Skin & midline fibrous septum of scrotum
Action Corrugates scrotal skin
Nerve Sympathetic fibres in genital br (L2) of genitofemoral N (L1,2)

DEEP TRANSVERSE PERINEI
Arises Med aspect of ischiopublic ramus & body of ischium within deep perineal pouch
Inserts Midline raphe & perineal body
Action Fixes perineal body & supports pelvic viscera
Nerve Perineal br of pudendal N (S2,3,4)

DELTOID
Arises Lat third of clavicle, acromion, spine of scapula to deltoid tubercle
Inserts Middle of lat surface of humerus (deltoid tuberosity)
Action Abducts arm, ant fibres flex & med rotate, post fibres extend & lat rotate
Nerve Axillary N (C5,6) (from post cord)

DEPRESSOR ANGULI ORIS
Arises Outer surface of mandible inf to mental foramen
Inserts Modiolus at angle of mouth
Action Depresses & draws angle of mouth laterally
Nerve Mandibular br of facial N (VII)

DEPRESSOR LABII INFERIORIS
Arises Outer surface of mandible inferomedial to mental foramen
Inserts Skin of lower lip
Action Depresses & draws lower lip laterally
Nerve Mandibular br of facial N (VII)

DIAPHRAGM
Arises Vertebral: crura from bodies of L1,2 (left), L1–3 (right), together giving median arcuate lig. Costal: med & lat arcuate ligs, inner aspect of lower six ribs and costal cartilages
Sternal: two slips from post aspect of xiphoid
Inserts Trefoil central tendon
Action Inspiration & assists in raising intra-abdominal pressure
Nerve Phrenic N (motor) (C3,4,5). Sensory: phrenic, intercostals (6–12) & upper two lumbar N roots

DIGASTRIC
Arises Ant belly: digastric fossa on post surface of symphysis menti. Post belly: base of med aspect of mastoid process
Inserts Fibrous loop to lesser cornu of hyoid bone
Action Elevates hyoid bone. Aids swallowing & depresses mandible
Nerve Ant belly: mylohyoid N (Vc). Post belly: facial N (VII) before it enters parotid gland

ERECTOR SPINAE—ILIOCOSTOCERVICALIS
Arises Post angles of ribs
Inserts Transverse processes above & below
Action Extends & lat flexes spine
Nerve Post primary rami
Notes Divided into iliocostalis—lumborum, thoracis & cervicalis

ERECTOR SPINAE—LONGISSIMUS
Arises Transverse processes
Inserts Transverse processes several levels above

Action Extends spine
Nerve Post primary rami
Notes Divided into longissimus—thoracis, cervicis & capitis

ERECTOR SPINAE—SPINALIS

Arises Spinous processes
Inserts Spinous processes six levels above
Action Lat flexion of spine
Nerve Post primary rami
Notes Divided into spinalis—thoracis, cervicis & capitis

EXTENSOR CARPI RADIALIS BREVIS

Arises Common extensor origin on ant aspect of lat epicondyle of humerus
Inserts Post base of 3rd MC
Action Extends & abducts hand at wrist
Nerve Post interosseous N (C7,8)

EXTENSOR CARPI RADIALIS LONGUS

Arises Lower third of lat supracondylar ridge of humerus & lat intermuscular septum
Inserts Post base of 2nd MC
Action Extends & abducts hand at wrist
Nerve Radial N (C6,7)

EXTENSOR CARPI ULNARIS

Arises Common extensor origin on ant aspect of lat epicondyle of humerus
Inserts Base of 5th MC via groove by ulnar styloid
Action Extends & adducts hand at wrist
Nerve Post interosseous N (C7,8)

EXTENSOR DIGITI MINIMI (hand)

Arises Common extensor origin on ant aspect of lat epicondyle of humerus
Inserts Extensor expansion of little finger—usually two tendons which are joined by a slip from extensor digitorum at MCP jnt
Action Extends all jnts of little finger
Nerve Post interosseous N (C7,8)

EXTENSOR DIGITORUM (hand)

Arises Common extensor origin on ant aspect of lat epicondyle of humerus
Inserts Extensor expansion to all phalanges of four fingers by four tendons. Tendons 3 & 4 usually fuse & little finger just receives a slip
Action Extends all jnts of fingers
Nerve Post interosseous N (C7,8)

EXTENSOR DIGITORUM BREVIS (foot)

Arises Sup surface of ant calcaneus
Inserts Four tendons into prox phalanx of big toe & long extensor tendons to toes 2, 3 and 4
Action Extends toes when foot fully dorsiflexed
Nerve Deep fibular (peroneal) N (L5,S1)
Notes Med one of four tendons could be regarded as extensor hallucis brevis

EXTENSOR DIGITORUM LONGUS (foot)

Arises Upper two-thirds of ant shaft of fibula, interosseous membrane & sup tibiofibular jnt
Inserts Extensor expansion of lat four toes
Action Extends toes & extends foot at ankle
Nerve Deep fibular (peroneal) N (L5,S1)

EXTENSOR HALLUCIS LONGUS

Arises Middle half of ant shaft of fibula
Inserts Dorsal base of distal phalanx of great toe
Action Extends big toe & foot. Inverts foot & tightens subtalar jnts
Nerve Deep fibular (peroneal) N (L5,S1)

EXTENSOR INDICIS

Arises Lower post shaft of ulna (below extensor pollicis longus) & adjacent interosseous membrane
Inserts Extensor expansion of index finger (tendon lies on ulnar side of extensor digitorum tendon)

Action Extends all jnts of index finger
Nerve Post interosseous N (C7,8)

EXTENSOR POLLICIS BREVIS
Arises Lower third of post shaft of radius & adjacent interosseous membrane
Inserts Over tendons of radial extensors & brachioradialis to base of prox phalanx of thumb
Action Extends MCP jnt of thumb
Nerve Post interosseous N (C7,8)
Notes Forms radial side of snuff box

EXTENSOR POLLICIS LONGUS
Arises Middle third of post ulna (below abductor pollicis longus) & adjacent interosseous membrane
Inserts Base of distal phalanx of thumb via Lister's tubercle (dorsal tubercle of radius)
Action Extends IP & MCP jnts of thumb
Nerve Post interosseous N (C7,8)
Notes Forms ulnar side of snuff box

EXTERNAL OBLIQUE ABDOMINIS
Arises Ant angles of lower eight ribs
Inserts Outer ant half of iliac crest, inguinal lig, pubic tubercle & crest, & aponeurosis of ant rectus sheath, linea alba & xiphisternum
Action Supports abdominal wall, assists forced expiration, aids raising intra-abdominal pressure &, with muscles of opposite side, abducts & rotates trunk
Nerve Ant primary rami (T7–12)
Notes Interdigitates with four slips of serratus anterior & four of latissimus dorsi

FIBULARIS (PERONEUS) BREVIS
Arises Lower two-thirds lat shaft of fibula
Inserts Tuberosity of base of 5th MT
Action Plantar flexes & everts foot. Supports lat longitudinal arch
Nerve Superficial fibular (peroneal) N (L5,S1)

FIBULARIS (PERONEUS) LONGUS
Arises Upper two-thirds of lat shaft of fibula, head of fibula & sup tibiofibular jnt
Inserts Plantar aspect of base of 1st MT & med cuneiform, passing deep to long plantar lig
Action Plantar flexes & everts foot. Supports lat longitudinal & transverse arches
Nerve Superficial fibular (peroneal) N (L5,S1)

FIBULARIS (PERONEUS) TERTIUS
Arises Third quarter of ant shaft of fibula
Inserts Dorsal shaft & base of 5th MT
Action Extends & everts foot
Nerve Deep fibular (peroneal) N (L5,S1)

FLEXOR CARPI RADIALIS
Arises Common flexor origin of med epicondyle of humerus
Inserts Bases of 2nd & 3rd MCs via groove in trapezium
Action Flexes & abducts wrist
Nerve Median N (C6,7) (from med & lat cords)

FLEXOR CARPI ULNARIS
Arises Humeral head: common flexor origin of med epicondyle. Ulnar head: aponeurosis from med olecranon & upper three-quarters subcutaneous border of ulna
Inserts Pisiform, hook of hamate, base of 5th MC via pisohamate & pisometacarpal ligs
Action Flexes & adducts wrist. Fixes pisiform during action of hypothenar muscles
Nerve Ulnar N (C7,8,T1) (by communication from lat cord)
Notes Ulnar N passes between two heads

FLEXOR DIGITI MINIMI BREVIS (foot)
Arises Base of 5th MT & sheath of fibularis (peroneus) longus
Inserts Lat side of base of prox phalanx of little toe
Action Flexes MTP jnt of little toe

Nerve Superficial br of lat plantar N (S2,3)
Notes A few muscle fibres to distal half of plantar surface of 5th MT represent opponens digiti minimi

FLEXOR DIGITI MINIMI BREVIS (hand)

Arises Flexor retinaculum & hook of hamate
Inserts Ulnar side of base of prox phalanx of little finger
Action Flexes MCP jnt of little finger
Nerve Deep br of ulnar N (C8,T1)

FLEXOR DIGITORUM ACCESSORIUS (QUADRATUS PLANTAE) (foot)

Arises Med & lat sides of calcaneus
Inserts Tendons of flexor digitorum longus
Action Assists flexor digitorum longus to flex lat four toes, especially when ankle is plantar flexed
Nerve Lat plantar N (S2,3)

8

FLEXOR DIGITORUM BREVIS (foot)

Arises Med process of post calcaneal tuberosity
Inserts Four tendons to four lat toes to both sides of middle phalanx. Tendons of flexor digitorum longus pass through them
Action Flexes lat four toes. Supports med & lat longitudinal arches
Nerve Med plantar N (S1,2)

FLEXOR DIGITORUM LONGUS (foot)

Arises Post shaft of tibia below soleal line & by broad aponeurosis from fibula
Inserts Base of distal phalanges of lat four toes
Action Flexes distal phalanges of lat four toes & foot at ankle. Supports lat longitudinal arch
Nerve Tibial N (S2,3)
Notes Med two tendons receive slips from flexor hallucis longus & all four receive insertion of flexor accessorius & each gives a lumbrical

FLEXOR DIGITORUM PROFUNDUS (hand)

Arises Upper three-quarters of ant & med surface of ulna as far round as subcutaneous border & narrow strip of interosseous membrane
Inserts Distal phalanges of med four fingers. Tendon to index finger separates early
Action Flexes distal IP jnts, then secondarily flexes prox IP & MCP jnts & wrist
Nerve Median N (ant interosseous) (C8,T1) ulnar N (C8,T1)
Notes Nerve supply as above in 60%. In 40% it is a 3:1 ratio either way

FLEXOR DIGITORUM SUPERFICIALIS (hand)

Arises Humeral head: common flexor origin of med epicondyle of humerus, med lig of elbow. Ulnar head: sublime tubercle (med border of coronoid process) & fibrous arch. Radial head: whole length of ant oblique line
Inserts Tendons split to insert onto sides of middle phalanges of med four fingers
Action Flexes prox IP jnts & secondarily MCP jnts & wrist
Nerve Median N (C7,8,T1) (from med & lat cords)
Notes Median N applied to under surface of muscle

FLEXOR HALLUCIS BREVIS

Arises Cuboid, lat cuneiform & tibialis posterior insertion over the two remaining cuneiforms
Inserts Med tendon to med side of base of prox phalanx of big toe. Lat tendon to lat side of same, both via sesamoids
Action Flexes MTP jnt of big toe. Supports med longitudinal arch
Nerve Med plantar N (S1,2)

FLEXOR HALLUCIS LONGUS

Arises Lower two-thirds of post fibula between median crest & post border,

lower intermuscular septum & interosseous membrane
Inserts Base of distal phalanx of big toe & slips to med two tendons of flexor digitorum longus
Action Flexes distal phalanx of big toe, flexes foot at ankle, supports med logitudinal arch
Nerve Tibial N (S2,3)

FLEXOR POLLICIS BREVIS
Arises Superficial head: flexor retinaculum & tubercle of trapezium. Deep head: capitate & trapezoid
Inserts Base of prox phalanx of thumb (via radial sesamoid)
Action Flexes MCP jnt of thumb
Nerve Recurrent (muscular) br of median N (C8,T1) (may also be from deep br of ulnar N (C8,T1))

FLEXOR POLLICIS LONGUS
Arises Ant surface of radius below ant oblique line & adjacent interosseous membrane
Inserts Base of distal phalanx of thumb
Action Flexes distal phalanx of thumb
Nerve Ant interosseous N (C7,8)

GASTROCNEMIUS
Arises Lat head: post surface of lat condyle of femur & highest of three facets on lat condyle. Med head: post surface of femur above med condyle
Inserts Tendo calcaneus to middle of three facets on post aspect of calcaneus
Action Plantar flexes foot. Flexes knee
Nerve Tibial N (S1,2)
Notes Main propulsive force for jumping

GEMELLUS INFERIOR
Arises Upper border of ischial tuberosity
Inserts Middle part of med aspect of greater trochanter of femur
Action Lat rotates & stabilises hip
Nerve N to quadratus femoris (L4,5,S1)

GEMELLUS SUPERIOR
Arises Spine of ischium
Inserts Middle part of med aspect of greater trochanter of femur
Action Lat rotates & stabilises hip
Nerve N to obturator internus (L5,S1,2)

GENIOGLOSSUS
Arises Sup mental spine on post surface of symphysis menti
Inserts Central mass of tongue & mucous membrane
Action Protracts tongue
Nerve Hypoglossal N (XII)

GENIOHYOID
Arises Inf mental spine on post surface of symphysis menti
Inserts Sup border of body of hyoid bone
Action Elevates & protracts hyoid bone. Depresses mandible
Nerve C1 fibres carried by hypoglossal N (XII)

GLUTEUS MAXIMUS
Arises Outer surface of ilium behind post gluteal line & post third of iliac crest, lumbar fascia, lat mass of sacrum, sacrotuberous lig & coccyx
Inserts Deepest quarter into gluteal tuberosity of femur, remaining three-quarters into iliotibial tract (ant surface of lat condyle of tibia)
Action Extends & lat rotates hip. Maintains knee extended via iliotibial tract
Nerve Inf gluteal N (L5,S1,2)
Notes Largest muscle in body

GLUTEUS MEDIUS
Arises Outer surface of ilium between post & ant gluteal lines
Inserts Posterolateral surface of greater trochanter of femur
Action Abducts & med rotates hip. Tilts pelvis on walking
Nerve Sup gluteal N (L4,5,S1)

GLUTEUS MINIMUS
Arises Outer surface of ilium between ant & inf gluteal lines
Inserts Ant surface of greater trochanter of femur
Action Abducts & med rotates hip. Tilts pelvis on walking
Nerve Sup gluteal N (L4,5,S1)

GRACILIS
Arises Outer surface of ischiopubic ramus
Inserts Upper med shaft of tibia behind sartorius
Action Adducts hip. Flexes knee & med rotates flexed knee
Nerve Ant div of obturator N (L2,3)

HYOGLOSSUS (& CHONDROGLOSSUS)
Arises Sup border of greater cornu of hyoid bone
Inserts Lat side of tongue
Action Depresses tongue
Nerve Hypoglossal N (XII)

ILIACUS
Arises Iliac fossa within abdomen
Inserts Lowermost surface of lesser trochanter of femur
Action Flexes hip
Nerve Femoral N in abdomen (L2,3)

INFERIOR OBLIQUE (see also obliquus capitis inferior)
Arises Orbital surface of maxilla behind orbital margin on med side
Inserts Post/inf quadrant of sclera behind equator of eyeball on lat side
Action Elevates eye in adduction. Extorts eye in abduction
Nerve Inf div of oculomotor N (III)

INFERIOR RECTUS
Arises Inf tendinous ring within orbit
Inserts Inf sclera ant to equator of eyeball
Action Depresses eye. Extorts eye in adduction
Nerve Inf div of oculomotor N (III)

INFRASPINATUS
Arises Med three-quarters of infraspinous fossa of scapula & fibrous intermuscular septa
Inserts Middle facet of greater tuberosity of humerus & capsule of shoulder jnt
Action Lat rotates arm & stabilises shoulder jnt
Nerve Suprascapular N (C5,6) (from upper trunk)
Notes Bursa under tendon over glenoid angle. Tendon forms part of rotator cuff

INTERCOSTALS EXTERNAL
Arises Inf border of ribs as far forwards as costochondral junctions. Beyond this is ant (ext) intercostal membrane
Inserts Sup border of ribs below, passing obliquely downwards & forwards
Action Fix intercostal spaces during respiration. Aids forced respiration by elevating ribs
Nerve Muscular collateral brs of intercostal Ns

INTERCOSTALS INNERMOST
Arises Int lat aspect of ribs above & below
Inserts Int aspect of ribs above & below
Action Fix intercostal spaces during respiration
Nerve Muscular collateral brs of intercostal Ns
Notes Innermost intercostals are one of three muscles that make up inner layer of thoracic wall muscles. Others are transversus thoracis (ant) & subcostalis (post)

INTERCOSTALS INTERNAL
Arises Inf border of ribs as far back as post angles. Beyond this is post (int) intercostal membrane

Inserts Sup border of ribs below, passing obliquely downwards & backwards
Action Fix intercostal spaces during respiration. Aids forced inspiration by elevating ribs
Nerve Muscular collateral brs of intercostal Ns

INTERNAL OBLIQUE ABDOMINIS

Arises Lumbar fascia, ant two-thirds of iliac crest & lat two-thirds of inguinal lig
Inserts Costal margin (ribs & costal cartilages), aponeurosis of rectus sheath (ant & post), conjoint tendon to pubic crest & pectineal line
Action Supports abdominal wall, assists forced respiration, aids raising intra-abdominal pressure &, with muscles of other side, abducts & rotates trunk. Conjoint tendon supports post wall of inguinal canal
Nerve Ant primary rami (T7–12) (conjoint tendon ilio-inguinal N (L1))

INTEROSSEI—DORSAL OF FOOT (4)

Arises Bipennate from inner aspects of shafts of all MTs
Inserts Bases of prox phalanges & dorsal extensor expansions of med side of 2nd toe & lat sides of 2nd, 3rd & 4th toes
Action Abduct 2nd, 3rd & 4th toes from axis of 2nd toe. Assist lumbricals in extending IP jnts whilst flexing MTP jnts
Nerve Lat plantar N (1–3: deep br; 4: superficial br) (S2,3)

INTEROSSEI—DORSAL OF HAND (4)

Arises Bipennate from inner aspects of shafts of all MCs
Inserts Prox phalanges & dorsal extensor expansion on radial side of index & middle fingers & ulnar side of middle & ring fingers
Action Abduct index, middle & ring fingers from axis of middle finger. Flex MCP jnt whilst extending IP jnts
Nerve Deep br of ulnar N (C8,T1)

INTEROSSEI—PALMAR OF HAND (3)

Arises Ant shafts of 2, 4, 5 MCs (unipennate)
Inserts Prox phalanges & dorsal extensor expansion on ulnar side of index & radial side of ring & little fingers
Action Adduct index, ring & little fingers to axis of middle finger. Flex MCP jnt whilst extending IP jnts
Nerve Deep br of ulnar N (C8,T1)

INTEROSSEI—PLANTAR OF FOOT (3)

Arises Inferomedial shafts of 3rd, 4th & 5th MTs (single heads)
Inserts Med sides of bases of prox phalanges with slips to dorsal extensor expansions of 3rd, 4th & 5th toes
Action Adduct 3rd, 4th & 5th toes to axis of 2nd toe. Assist lumbricals in extending IP jnts whilst flexing MTP jnts
Nerve Deep br of lat plantar N (S2,3)

INTERSPINALES

Arises Spinous processes
Inserts Spinous processes one above
Action Extension of spine
Nerve Post primary rami

INTERTRANSVERSARII

Arises Transverse processes
Inserts Transverse processes one above
Action Lat flexes spine
Nerve Post primary rami

INTRINSIC MUSCLE OF TONGUE

Arises Sup & inf longitudinal, transverse & vertical elements
Inserts Mucous membrane, septum & other muscles of tongue
Action Alter shape of tongue & so aid mastication, speech & swallowing
Nerve Hypoglossal N (XII)

ISCHIOCAVERNOSUS

Arises Med aspect of ischium & ischiopubic ramus

Inserts Inferolateral aponeurosis over crura of penis/clitoris
Action Stabilises erect penis
Nerve Perineal br of pudendal N (S2,3,4)

LATERAL CRICOARYTENOID
Arises Lat aspect of arch of cricoid cartilage
Inserts Muscular process of arytenoid cartilage
Action Adducts & med rotates arytenoid cartilage (closes rima glottidis)
Nerve Recurrent laryngeal br of vagus N (X)

LATERAL PTERYGOID
Arises Upper head: infratemporal surface of sphenoid bone. Lower head: lat surface of lat pterygoid plate
Inserts Pterygoid fovea below condylar process of mandible & intra-articular disc & capsule of temporomandibular jnt
Action Protrudes mandible & opens mouth by pulling condyle & disc forwards
Nerve Ns to lat pterygoid (ant div of mandibular N (Vc))

LATERAL RECTUS
Arises Lat tendinous ring within orbit
Inserts Lat sclera ant to equator of eyeball
Action Abducts eye
Nerve Abducent N (VI)

LATISSIMUS DORSI
Arises All thoracic spines & supraspinous ligs from T7 downwards & lumbar & sacral spines via lumbar fascia, post third iliac crest, last four ribs (interdigitating with ext oblique abdominis) & inf angle of scapula
Inserts Floor of bicipital groove of humerus after spiraling around teres major
Action Extends, adducts & med rotates arm. Costal attachment helps with deep inspiration & forced expiration
Nerve Thoracodorsal N (C6,7,8) (from post cord)

LEVATOR ANGULI ORIS
Arises Ant surface of maxilla below infra-orbital foramen
Inserts Outer end of upper lip & modiolus
Action Elevates angle of mouth
Nerve Buccal br of facial N (VII)

LEVATOR ANI—COCCYGEUS
Arises Sacrospinous lig & ischial spine
Inserts Anococcygeal body & coccyx
Action Supports pelvic viscera
Nerve Ant primary rami (perineal brs) of S4,5

LEVATOR ANI—ILIOCOCCYGEUS
Arises Post half of fascial line over obturator internus & ischial spine
Inserts Anococcygeal body
Action Supports pelvic viscera
Nerve Ant primary rami (perineal brs) of S3,4

LEVATOR ANI—PUBOCOCCYGEUS
Arises Post surface of pubis & ant half of fascial line over obturator internus
Inserts Anococcygeal body
Action Supports pelvic viscera
Nerve Ant primary rami (perineal brs) of S3,4

LEVATOR ANI—PUBORECTALIS
Arises Post surface of pubis
Inserts Midline sling post to rectum
Action Supports & aids continence of rectum by maintaining anorectal angle
Nerve Ant primary rami (perineal brs) of S3,4

LEVATOR ANI—PUBOVAGINALIS (LEVATOR PROSTATAE)
Arises Post surface of pubis
Inserts Midline raphe post to vagina/prostate
Action Supports ant pelvic viscera
Nerve Ant primary rami (perineal brs) of S3,4

LEVATOR LABII SUPERIORIS
Arises Med infra-orbital margin
Inserts Skin & muscle of upper lip
Action Elevates & everts upper lip
Nerve Buccal br of facial N (VII)

LEVATOR LABII SUPERIORIS ALAEQUE NASI
Arises Upper frontal process of maxilla
Inserts Skin of lat nostril & upper lip
Action Dilates nostril & elevates upper lip
Nerve Buccal br of facial N (VII)

LEVATOR PALPEBRAE SUPERIORIS
Arises Inf aspect of lesser wing of sphenoid bone just above tendinous ring
Inserts Sup tarsal plate & skin of upper eyelid
Action Elevates & retracts upper eyelid
Nerve Sup div of oculomotor N (III) & sympathetic to smooth muscle portion

LEVATOR SCAPULAE
Arises Post tubercles of transverse processes of C1–4
Inserts Upper part of med border of scapula
Action Raises med border of scapula
Nerve Ant primary rami of C3 & C4 & dorsal scapular N (C5)

LEVATOR VELI PALATINI
Arises Within pharynx at apex of inf surface of petrous temporal bone & med rim of auditory tube
Inserts Palatine aponeurosis
Action Elevates, retracts & lat deviates soft palate. May open auditory tube on swallowing
Nerve Pharyngeal plexus (IX, X, sympathetic) via pharyngeal br of vagus N (X) with its motor fibres from cranial accessory N (XI)

LEVATORES COSTARUM
Arises Transverse processes C7 to T11
Inserts Post surface & angle of rib below
Action Elevates ribs
Nerve Post primary rami

LONGUS CAPITIS
Arises Ant tubercles of transverse processes of C3–6
Inserts Ant basilar occipital bone
Action Flexes cervical spine & atlanto-occipital jnt
Nerve Ant primary rami of C1–3

LONGUS COLLI
Arises Ant body of T1–3, ant tubercles of transverse processes of C3–7
Inserts Ant arch of atlas (C1) & bodies of C2–4
Action Flexes & rotates cervical spine
Nerve Ant primary rami of C2–6

LUMBRICALS OF FOOT (4)
Arises Lat 3: bipennate origin from cleft between the four tendons of flexor digitorum longus. Med 1: unipennate origin from med aspect of 1st tendon
Inserts Dorsal extensor expansion
Action Extend toes at IP jnts & flex MTP jnts
Nerve First: med plantar N (S2,3). 2–4: deep br of lat plantar N (S2,3)

LUMBRICALS OF HAND (4)
Arises Four tendons of flexor digitorum profundus. Radial 2: radial side only (unipennate). Ulnar 2: cleft between tendons (bipennate)
Inserts Extensor expansion (dorsum of prox phalanx) of fingers 2–5 radial side
Action Flex MCP jnts & extend IP jnts of fingers
Nerve Lat 2: median N (C8,T1). Med 2: deep br of ulnar N (C8,T1)
Notes 60% have nerve supply as above. 40% have 3:1 or 1:3

MASSETER
Arises Ant two-thirds of zygomatic arch & zygomatic process of maxilla

Inserts Lat surface of angle & lower ramus of mandible
Action Elevates mandible (enables forced closure of mouth)
Nerve Ant div of mandibular N (Vc)

MEDIAL PTERYGOID
Arises Deep head. Med side of lat pterygoid plate & fossa between med & lat plates. Superficial head: tuberosity of maxilla & pyramidal process of palatine bone
Inserts Med aspect of angle of mandible
Action Elevates, protracts & lat displaces mandible to opposite side for chewing
Nerve N to medial pterygoid (main trunk of mandibular N (Vc))

MEDIAL RECTUS
Arises Med tendinous ring within orbit
Inserts Med sclera ant to equator of eyeball
Action Adducts eye
Nerve Inf div of oculomotor N (III)

8

MENTALIS
Arises Incisive fossa on ant aspect of mandible
Inserts Skin of chin
Action Elevates & wrinkles skin of chin & protrudes lower lip
Nerve Mandibular br of facial N (VII)

MUSCULUS UVULAE
Arises Post border of hard palate
Inserts Palatine aponeurosis
Action Shapes uvula
Nerve Pharyngeal plexus (IX, X, sympathetic) via pharyngeal br of vagus N (X) with its motor fibres from cranial accessory N (XI)

MYLOHYOID
Arises Mylohyoid line on int aspect of mandible
Inserts Ant three-quarters: midline raphe. Post quarter: sup border of body of hyoid bone
Action Elevates hyoid bone, supports & raises floor of mouth. Aids in mastication & swallowing
Nerve Mylohyoid N (Vc)

NASALIS (COMPRESSOR & DILATOR)
Arises Frontal process of maxilla
Inserts Nasal aponeurosis
Action Opens & closes nostrils, particularly in forced respiration
Nerve Buccal br of facial N (VII)
Notes Part of dilator nasalis is depressor septi from maxilla above central incisor to mobile part of nasal septum

OBLIQUE ARYTENOID
Arises Muscular process of arytenoid cartilage
Inserts Sup pole of opposite arytenoid cartilage
Action Adducts arytenoid cartilages (closes rima glottidis)
Nerve Recurrent laryngeal br of vagus N (X)
Notex It extends into aryepiglottic fold as aryepiglotticus

OBLIQUUS CAPITIS INFERIOR
Arises Spinous process of axis (C2)
Inserts Lat mass of atlas (C1)
Action Rotates atlanto-axial jnt
Nerve Suboccipital N (post primary ramus of C1)

OBLIQUUS CAPITIS SUPERIOR
Arises Lat mass of atlas (C1)
Inserts Lat half inf nuchal line
Action Lat flexes atlanto-occipital jnt
Nerve Suboccipital N (post primary ramus of C1)

OBTURATOR EXTERNUS
Arises Outer obturator membrane, rim of pubis & ischium bordering it
Inserts Trochanteric fossa on med surface of greater trochanter

Action Lat rotates hip
Nerve Post div of obturator N (L3,4)

OBTURATOR INTERNUS
Arises Inner surface of obturator membrane & rim of pubis & ischium bordering membrane
Inserts Middle part of med aspect of greater trochanter of femur
Action Lat rotates & stabilises hip
Nerve N to obturator internus (L5,S1,2)

OCCIPITOFRONTALIS
Arises Occipital: highest nuchal line & mastoid process. Frontal: sup fibres of upper facial muscles
Inserts Galeal aponeurosis
Action Wrinkles forehead & fixes galeal aponeurosis
Nerve Post auricular & temporal brs of facial N (VII)

OMOHYOID
Arises Suprascapular lig & adjacent scapula
Inserts Inf border of body of hyoid bone
Action Depresses hyoid bone & hence larynx
Nerve Ansa cervicalis N (C1,2,3)
Notes Tendon between two bellies through sling behind sternocleidomastoid

OPPONENS DIGITI MINIMI (hand)
Arises Flexor retinaculum & hook of hamate
Inserts Ulnar border of shaft of 5th MC
Action Opposes (flexes & lat rotates) CMC jnt of little finger
Nerve Deep br of ulnar N (C8,T1)

OPPONENS POLLICIS
Arises Flexor retinaculum & tubercle of trapezium
Inserts Whole of radial border of 1st MC
Action Opposes (med rotates & flexes) CMC jnt of thumb
Nerve Recurrent (muscular) br of median N (C8,T1) (may also be from deep br of ulnar N (C8,T1))

ORBICULARIS OCULI
Arises Med orbital margin & lacrimal fascia behind lacrimal sac (orbital, palpebral & lacrimal parts)
Inserts Lat palpebral raphe
Action Closes eyelids, aids passage & drainage of tears
Nerve Temporal & zygomatic brs of facial N (VII)

ORBICULARIS ORIS
Arises Near midline on ant surface of maxilla & mandible & modiolus at angle of mouth
Inserts Mucous membrane of margin of lips & raphe with buccinator at modiolus
Action Narrows orifice of mouth, purses lips & puckers lip edges
Nerve Buccal br of facial N (VII)
Notes Accessory parts are incisivus labii superioris & inferioris

PALATOGLOSSUS
Arises Palatine aponeurosis
Inserts Posterolateral tongue
Action Elevates post tongue & closes oropharyngeal isthmus & aids initiation of swallowing
Nerve Pharyngeal plexus (IX, X, sympathetic) via pharyngeal br of vagus N (X) with its motor fibres from cranial accessory N (XI)
Notes Forms palatoglossal arch

PALATOPHARYNGEUS
Arises Palatine aponeurosis & post margin of hard palate
Inserts Upper border of thyroid cartilage & blends with constrictor fibres. Upper fibres interdigitate with opposite side to give Passavant's ridge

Action Elevates pharynx & larynx. Passavant's muscle closes nasopharyngeal isthmus in swallowing
Nerve Pharyngeal plexus (IX, X, sympathetic) via pharyngeal br of vagus N (X) with its motor fibres from cranial accessory N (XI)
Notes Forms palatoglopharyngeal arch

PALMARIS BREVIS
Arises Flexor retinaculum & palmar aponeurosis
Inserts Skin of palm into dermis
Action Steadies & corrugates skin of palm to help with grip
Nerve Superficial br of ulnar N (C8,T1)
Notes Only muscle supplied by this br of ulnar N

PALMARIS LONGUS
Arises Common flexor origin of med epicondyle of humerus
Inserts Flexor retinaculum & palmar aponeurosis
Action Flexes wrist & tenses palmar aponeurosis
Nerve Median N (C7,8) (from med & lat cords)
Notes Absent in 13%

PECTINEUS
Arises Pectineal line of pubis & narrow area of sup pubic ramus below it
Inserts A vertical line between spiral line & gluteal crest below lesser trochanter of femur
Action Flexes & adducts hip
Nerve Ant div of femoral N (L2,3). Occasional twig from obturator N (ant div—L2,3)

PECTORALIS MAJOR
Arises Clavicular head—med half clavicle. Sternocostal head—lat manubrium & sternum, six upper costal cartilages & ext oblique aponeurosis
Inserts Lat lip of bicipital groove of humerus and ant lip of deltoid tuberosity
Action Clavicular head: flexes & adducts arm. Sternal head: adducts & med rotates arm. Accessory for inspiration
Nerve Med pectoral N (from med cord) & lat pectoral N (from lat cord) (C6,7,8)
Notes Muscle folds on itself so that clavicular fibres insert lowest. Sternal fibres are highest inserting into capsule of shoulder joint

PECTORALIS MINOR
Arises Ant aspect of 3, 4, 5 ribs
Inserts Med & upper surface of coracoid process of scapula
Action Elevates ribs if scapula fixed, protracts scapula (assists serratus anterior)
Nerve Med & lat pectoral Ns (C6,7,8) (from med & lat cords)
Notes Landmark for axillary art & cords of brachial plexus

PIRIFORMIS
Arises 2, 3, 4 costotransverse bars of ant sacrum between sacral foramina
Inserts Ant part of med aspect of greater trochanter of femur
Action Lat rotates & stabilises hip
Nerve Ant primary rami of S1,2
Notes Passes lat through greater sciatic foramen

PLANTARIS
Arises Lat supracondylar ridge of femur above lat head of gastrocnemius
Inserts Tendo calcaneus (med side, deep to gastrocnemius tendon)
Action Plantar flexes foot & flexes knee
Nerve Tibial N (S1,2)

PLATYSMA
Arises Skin over lower neck & upper lat chest
Inserts Inf border of mandible & skin over lower face & angle of mouth

Action Depresses & wrinkles skin of lower face & mouth. Aids forced depression of mandible
Nerve Cervical br of facial N (VII)

POPLITEUS
Arises Post shaft of tibia above soleal line & below tibial condyles
Inserts A facet on lat surface of lat condyle of femur postero-inferior to epicondyle. Tendon passes into capsule of knee and a few fibres attach to lat meniscus
Action Unlocks extended knee by lat rotation of femur on tibia. Pulls back lat meniscus
Nerve Tibial N (L4,5,S1)
Notes Popliteus bursa lies deep to tendon

POSTERIOR CRICOARYTENOID
Arises Post aspect of cricoid cartilage
Inserts Muscular process of arytenoid cartilage
Action Abducts & lat rotates arytenoid cartilage (opens rima glottidis)
Nerve Recurrent laryngeal br of vagus N (X)

PROCERUS
Arises Nasal bone & cartilages
Inserts Skin of med forehead
Action Wrinkles & 'frowns' forehead
Nerve Temporal br of facial N (VII)

PRONATOR QUADRATUS
Arises Lower quarter of anteromedial shaft of ulna
Inserts Lower quarter of anterolateral shaft of radius & some interosseous membrane
Action Pronates forearm & maintains ulna & radius opposed
Nerve Ant interosseous N (C8,T1)

PRONATOR TERES
Arises Humeral head: med epicondyle, med supracondylar ridge & med intermuscular septum. Ulnar head: med border of coronoid process
Inserts Just post to most prominent part of lat convexity of radius
Action Pronates forearm & flexes elbow
Nerve Median N (C6,7) (from lat & med cords)
Notes Median N passes between its two heads

PSOAS MAJOR
Arises Transverse processes of L1–5, bodies of T12–L5 & intervertebral discs below bodies of T12–L4
Inserts Middle surface of lesser trochanter of femur
Action Flexes hip
Nerve Ant primary rami of L1,2

PSOAS MINOR
Arises Bodies of T12 & L1 & intervening intervertebral disc
Inserts Pectineal line of pubis & fascia over iliopsoas
Action Weak flexor of trunk
Nerve Ant primary rami of L1

PYRAMIDALIS
Arises Public crest ant to origin of rectus abdominis
Inserts Lower linea alba
Action Reinforces lower rectus sheath
Nerve Subcostal N (T12)

QUADRATUS FEMORIS
Arises Lat border of ischial tuberosity
Inserts Quadrate tubercle of femur & a vertical line below this to the level of lesser trochanter
Action Lat rotates & stabilises hip
Nerve N to quadratus femoris (L4,5,S1)

QUADRATUS LUMBORUM
Arises Inf border of 12th rib
Inserts Apices of transverse processes of L1–4, iliolumbar lig & post third of iliac crest

Action Fixes 12th rib during respiration & lat flexes trunk
Nerve Ant primary rami (T12–L3)

RECTUS ABDOMINIS
Arises Pubic crest & pubic symphysis
Inserts 5, 6, 7 costal cartilages, med inf costal margin & post aspect of xiphoid
Action Flexes trunk, aids forced expiration & raises intra-abdominal pressure
Nerve Ant primary rami (T7–12)

RECTUS CAPITIS ANTERIOR
Arises Lat mass of atlas (C1)
Inserts Basilar occipital bone ant to occipital condyle
Action Flexes atlanto-occipital jnt
Nerve Ant primary rami of C1

RECTUS CAPITIS LATERALIS
Arises Lat mass of atlas (C1)
Inserts Jugular process of occipital bone
Action Lat flexes atlanto-occipital jnt
Nerve Ant primary rami of C1

8

RECTUS CAPITIS POSTERIOR MAJOR
Arises Spinous process of axis (C2)
Inserts Lat half of inf nuchal line
Action Extends & rotates atlanto-occipital jnt
Nerve Suboccipital N (post primary ramus C1)

RECTUS CAPITIS POSTERIOR MINOR
Arises Post process of atlas (C1)
Inserts Med half of inf nuchal line
Action Extends atlanto-occipital jnt
Nerve Suboccipital N (post primary ramus C1)

RECTUS FEMORIS (QUADRICEPS FEMORIS I)
Arises Straight head: ant inf iliac spine. Reflected head: ilium above acetabulum
Inserts Quadriceps tendon to patella, via ligamentum patellae into tubercle of tibia
Action Extends leg at knee. Flexes thigh at hip
Nerve Post div of femoral N (L2,3,4)

RHOMBOID MAJOR
Arises Spines of T2–T5 & supraspinous ligs
Inserts Lower half of posteromedial border of scapula, from angle to upper part of triangular area at base of scapular spine
Action Retracts scapula. Rotates scapula to rest position
Nerve Dorsal scapular N (C5) (from root)

RHOMBOID MINOR
Arises Lower ligamentum nuchae, spines of C7 & T1
Inserts Small area of posteromedial border of scapula at level of spine, below levator scapulae
Action Retracts scapula. Rotates scapula to rest position
Nerve Dorsal scapular N (C5) (from root)

RISORIUS
Arises Deep fascia of face & parotid
Inserts Modiolus & skin at angle of mouth
Action Retracts angle of mouth
Nerve Buccal br of facial N (VII)

SALPINGOPHARYNGEUS
Arises Inf cartilage & mucosa of pharyngeal orifice of auditory tube
Inserts Upper border of thyroid cartilage & inf constrictor muscle fibres
Action Elevates pharynx & larynx & aids swallowing. Opens auditory tube during swallowing
Nerve Pharyngeal plexus (IX, X, sympathetic) via pharyngeal br of vagus N (X) with its motor fibres from cranial accessory N (XI)

SARTORIUS
Arises Immediately below ant sup iliac spine

Inserts Upper med surface of shaft of tibia
Action Flexes, abducts, lat rotates thigh at hip. Flexes, med rotates leg at knee
Nerve Ant div of femoral N (L2,3)

SCALENUS ANTERIOR
Arises Ant tubercles of transverse processes of C3–6
Inserts Scalene tubercle on sup aspect of 1st rib
Action Accessory to inspiration. Lat flexion of neck when 1st rib fixed
Nerve Ant primary rami of C5,6

SCALENUS MEDIUS
Arises Post tubercles of transverse processes of C2–7
Inserts Sup aspect of 1st rib, ant to its tubercle
Action Accessory to inspiration
Nerve Ant primary rami of C3–8

SCALENUS MINIMUS
Arises Ant tubercle of transverse process of C7
Inserts Suprapleural membrane (Sibson's fascia)
Action Supports suprapleural membrane
Nerve Ant primary rami of C7

SCALENUS POSTERIOR
Arises Post tubercles of transverse processes C4–6
Inserts Posterolateral surface of 2nd rib
Action Accessory to inspiration
Nerve Ant primary rami of C6–8

SEMIMEMBRANOSUS
Arises Upper outer quadrant of post surface of ischial tuberosity
Inserts Med condyle of tibia below articular margin, fascia over popliteus & oblique popliteal lig
Action Flexes & med rotates knee. Extends hip
Nerve Tibial portion of sciatic N (L5,S1,2)

SEMITENDINOSUS
Arises Upper inner quadrant of post surface of ischial tuberosity
Inserts Upper med shaft of tibia behind gracilis
Action Flexes & med rotates knee. Extends hip
Nerve Tibial portion of sciatic N (L5,S1,2)

SERRATUS ANTERIOR
Arises Upper eight ribs & ant intercostal membranes from midclavicular line. Lower four interdigitating with external oblique
Inserts Inner med border scapula. 1 & 2: upper angle; 3 & 4: length of costal surface; 5–8: inf angle
Action Lat rotates & protracts scapula
Nerve Long thoracic N of Bell (C5,6,7) (from roots) slips from ribs 1 & 2: C5; 3 & 4: C6; 5–8: C7

SERRATUS POSTERIOR INFERIOR
Arises Spinous processes & supraspinous ligs of T11–L2
Inserts Post aspect of ribs 9–12
Action Assists forced expiration
Nerve Ant primary rami (T9–12)

SERRATUS POSTERIOR SUPERIOR
Arises Spinous processes & supraspinous lig of C7–T2
Inserts Post aspect of ribs 2–5
Action Assists forced inspiration
Nerve Ant primary rami (T2–5)

SOLEUS
Arises Soleal line & middle third of post border of tibia & upper quarter of post shaft of fibula including neck
Inserts Tendo calcaneus to middle of three facets on post surface of calcaneus
Action Plantar flexes foot (aids venous return)
Nerve Tibial N (S1,2)
Notes Main propulsive force for walking & running

SPHINCTER ANI (external)
Arises Circular anatomical sphincter
Inserts Deep, superficial & subcutaneous portions
Action Maintains continence of faeces
Nerve Inf rectal br of pudendal N (S2,3,4)

SPHINCTER URETHRAE
Arises Circular anatomical sphincter
Inserts Fuses with deep transverse perinei
Action Maintains continence of urine
Nerve Perineal br of pudendal N (S2,3,4)

SPLENIUS CAPITIS
Arises Lower lig nuchae, spinous processes & supraspinous ligs C7–T3
Inserts Lat occiput between sup & inf nuchal lines
Action Extends & rotates cervical spine
Nerve Post primary rami of C3,4

SPLENIUS CERVICIS
Arises Spinous processes & supraspinous ligs of T3–6
Inserts Post tubercles of transverse processes of C1–3
Action Extends & rotates cervical spine
Nerve Post primary rami of C5,6

STAPEDIUS
Arises The pyramid (post wall of middle ear)
Inserts Neck of stapes
Action Protects & critically damps ossicular chain
Nerve Facial N (VII), in middle ear

STERNOCLEIDOMASTOID
Arises Ant & sup manubrium & superomedial third of clavicle
Inserts Lat aspect of mastoid process & ant half of sup nuchal line
Action Acting alone it laterally flexes cervical spine & rotates head on neck to bring ipsilateral ear to ipsilateral shoulder. Acting together they protrude head & if head is fixed, aid respiration
Nerve Spinal root of accessory N (XI) (lat roots C1–5)
Notes Effectively four separate muscles

STERNOHYOID
Arises Sup lat post aspect of manubrium
Inserts Inf border of body of hyoid bone
Action Depresses hyoid bone & hence larynx
Nerve Ansa cervicalis N (C1,2,3)

STERNOTHYROID
Arises Med post aspect of manubrium
Inserts Oblique line of lamina of thyroid cartilage
Action Depresses larynx
Nerve Ansa cervicalis N (C1,2,3)

STYLOGLOSSUS
Arises Ant surface & apex of styloid process & upper quarter of stylohyoid lig
Inserts Superolateral sides of tongue
Action Retracts & elevates tongue, aids initiation of swallowing
Nerve Hypoglossal N (XII)

STYLOHYOID
Arises Base of styloid process
Inserts Base of greater cornu of hyoid bone
Action Elevates & retracts hyoid bone. Aids swallowing & elevates larynx
Nerve Facial N (VII) before it enters parotid gland

STYLOPHARYNGEUS
Arises Med aspect of styloid process
Inserts Posterolateral border of thyroid cartilage
Action Elevates larynx & pharynx. Aids swallowing
Nerve Muscular br of glossopharyngeal N (IX)

SUBCLAVIUS
Arises Costochondral junction of 1st rib

Inserts Subclavian groove on inf surface of middle third of clavicle
Action Depresses clavicle & steadies it during shoulder movements
Nerve N to subclavius (C5,6, upper trunk)

SUBCOSTALIS
Arises Int post aspects of lower six ribs
Inserts Int aspects of ribs two to three levels below
Action Depresses lower ribs
Nerve Muscular collateral brs of intercostal Ns
Notes Subcostalis is one of three muscles that make up inner layer of thoracic wall muscles. Others are innermost intercostals (lat) & transversus thoracis (ant)

SUBSCAPULARIS
Arises Med two-thirds of subscapular fossa
Inserts Lesser tuberosity of humerus, upper med lip of bicipital groove, capsule of shoulder jnt
Action Med rotates arm & stabilises shoulder jnt
Nerve Upper & lower subscapular Ns (C5,6) (from post cord)
Notes Subscapular bursa beneath tendon, usually connected with jnt. Tendon forms part of rotator cuff

SUPERFICIAL TRANSVERSE PERINEI
Arises Body of ischium
Inserts Perineal body
Action Fixes perineal body
Nerve Perineal br of pudendal N (S2,3,4)

SUPERIOR OBLIQUE (see also obliquus capitis superior)
Arises Body of sphenoid superomedial to tendinous ring
Inserts Post/sup quadrant of sclera behind equator of eyeball on lat side
Action Depresses eye in adduction. Intorts eye in abduction
Nerve Trochlear N (IV)
Notes Passes around trochlear sling on frontal bone

SUPERIOR RECTUS
Arises Sup tendinous ring within orbit
Inserts Sup sclera ant to equator of eyeball
Action Elevates eye. Intorts eye in adduction
Nerve Sup div of oculomotor N (III)

SUPINATOR
Arises Deep part (horizontal): supinator crest & fossa of ulna. Superficial part (downwards): lat epicondyle & lat lig of elbow & annular lig
Inserts Neck & shaft of radius, between ant & post oblique lines
Action Supinates forearm. Only acts alone when elbow extended
Nerve Post interosseous N (C6,7)
Notes Post interosseous N passes between its two heads

SUPRASPINATUS
Arises Med three-quarters of supraspinous fossa of scapula, upper surface of spine (bipennate)
Inserts Sup facet on greater tuberosity of humerus & capsule of shoulder jnt
Action Abducts arm & stabilises shoulder jnt
Nerve Suprascapular N (C5,6) (from upper trunk)
Notes Subacromial bursa lies above its tendon. Tendon forms part of rotator cuff

TEMPORALIS
Arises Temporal fossa between inf temporal line & infratemporal crest
Inserts Med & ant aspects of coronoid process of mandible
Action Elevates mandible & post fibres retract it
Nerve Deep temporal brs from ant div of mandibular N (Vc)

TEMPOROPARIETALIS
Arises Aponeurosis above ear
Inserts Galeal aponeurosis
Action Fixes galeal aponeurosis
Nerve Temporal br of facial N (VII)

TENSOR FASCIAE LATAE
Arises Outer surface of ant iliac crest between tubercle of the iliac crest & ant sup iliac spine
Inserts Iliotibial tract (ant surface of lat condyle of tibia)
Action Maintains knee extended (assists gluteus maximus) & abducts hip
Nerve Sup gluteal N (L4,5,S1)

TENSOR TYMPANI
Arises Cartilaginous & bony margins of auditory tube
Inserts Handle of malleus (via processus trochleariformis)
Action Protects & critically damps ossicular chain
Nerve Off N to med pterygoid (main trunk of mandibular N (Vc))

TENSOR VELI PALATINI
Arises Scaphoid fossa, med aspect of spine of sphenoid bone & lat cartilage of auditory tube
Inserts Palatine aponeurosis (via pulley of pterygoid hamulus)
Action Tenses soft palate prior to elevation. Opens auditory tube on swallowing
Nerve Off N to med pterygoid (main trunk of mandibular N (Vc))

TERES MAJOR
Arises Oval area (lower third) of lat side of inf angle of scapula below teres minor
Inserts Med lip of bicipital groove of humerus
Action Med rotates & adducts arm. Stabilises shoulder jnt
Nerve Lower subscapular N (C5,6,7) (from post cord)
Notes Functionally part of subscapularis

TERES MINOR
Arises Middle third lat border of scapula above teres major
Inserts Inf facet of greater tuberosity of humerus (below infraspinatus) & capsule of shoulder jnt
Action Lat rotates arm & stabilises shoulder jnt
Nerve Axillary N (C5,6) (from post cord)
Notes Functionally part of infraspinatus. Tendon forms part of rotator cuff

THYRO-ARYTENOID & VOCALIS
Arises Lower post surface of angle between laminae of thyroid cartilage
Inserts Vocal process of arytenoid cartilage
Action Shortens & relaxes vocal cords by approximating arytenoid cartilage to thyroid cartilage
Nerve Recurrent laryngeal br of vagus N (X)
Notes Vocalis is that part of thyro-arytenoid that inserts into vocal cord itself

THYRO-EPIGLOTTICUS
Arises Lower post surface of thyroid cartilage
Inserts Lat border of epiglottis
Action Widens additus of larynx
Nerve Recurrent laryngeal br of vagus N (X)

THYROHYOID
Arises Oblique line on lamina of thyroid cartilage
Inserts Inf border of body of hyoid bone
Action Elevates larynx or depresses hyoid bone
Nerve C1 fibres carried by hypoglossal N (XII)

TIBIALIS ANTERIOR
Arises Upper half of lat shaft of tibia & interosseous membrane
Inserts Inferomedial aspect of med cuneiform & base of 1st MT

Action Extends & inverts foot at ankle. Holds up med longitudinal arch of foot
Nerve Deep fibular (peroneal) N (L4,5)
Notes Inversion is at subtalar & mid tarsal joints

TIBIALIS POSTERIOR

Arises Upper half of post shaft of tibia & upper half of fibula between median crest & interosseous border, & interosseous membrane
Inserts Tuberosity of navicular bone & all tarsal bones (except talus) & bases of metatarsals 2–4
Action Plantar flexes & inverts foot. Supports med longitudinal arch of foot
Nerve Tibial N (L4,5)

TRANSVERSE ARYTENOID

Arises Post surface & muscular process of arytenoid cartilage
Inserts Corresponding surfaces of opposite cartilage
Action Adducts arytenoid cartilages (closes rima glottidis)
Nerve Recurrent laryngeal branch of vagus N (X)

TRANSVERSOSPINALIS—MULTIFIDUS

Arises Laminae of vertebra from sacrum to C2
Inserts Spinous processes two to three levels above
Action Extends spine
Nerve Post primary rami

TRANSVERSOSPINALIS—ROTATORES

Arises Transverse processes
Inserts Spinous processes one above
Action Rotates spine
Nerve Post primary rami
Notes Two types—thoracis & cervicis et lumborum

TRANSVERSOSPINALIS—SEMISPINALIS

Arises Transverse processes
Inserts Spinous processes six levels above
Action Extends & lat flexes spine
Nerve Post primary rami
Notes Three types—thoracis, cervicis & capitis

TRANSVERSUS ABDOMINIS

Arises Costal margin (ribs & costal cartilages), lumbar fascia, ant two-thirds of iliac crest & lat half of inguinal lig
Inserts Aponeurosis of post & ant rectus sheath & conjoint tendon to pubic crest & pectineal line
Action Supports abdominal wall, aids forced expiration & raising intra-abdominal pressure. Conjoint tendon supports post wall of inguinal canal
Nerve Ant primary rami (T7–12). Conjoint tendon: ilio-inguinal N (L1)

TRANSVERSUS THORACIS (STERNOCOSTALIS)

Arises Lower third of inner aspect of sternum & lower three costosternal junctions
Inserts Second to sixth costal cartilages
Action Depresses upper ribs
Nerve Muscular collateral brs of intercostal Ns
Notes Transversus thoracis is one of three muscles that make up inner layer of thoracic wall muscles. Others are innermost intercostals (lat) & subcostals (post)

TRAPEZIUS

Arises Med third sup nuchal line, lig nuchae, spinous processes & supraspinous ligs to T12
Inserts Upper fibres to lat third of post border of clavicle; med acromion & lat spine of scapula. Lower fibres to med end of spine of scapula as far as deltoid tubercle
Action Elevates & retracts scapula. Rotates it during abduction of arm. If scapula is fixed, extends & lat flexes neck

Nerve Spinal root of accessory N (XI) (lat roots, C1–5) (spinal nerves C3 & C4 for proprioception)

TRICEPS

Arises Long head: infraglenoid tubercle of scapula. Lat head: upper half post humerus (linear origin). Med head: lower half post humerus inferomedial to spiral groove & both intermuscular septa

Inserts Post part of upper surface of olecranon process of ulna & post capsule

Action Extends elbow. Long head stabilises shoulder jnt. Med head retracts capsule of elbow jnt on extension

Nerve Radial N (C6,7,8) (from post cord), four brs

VASTUS INTERMEDIUS (QUADRICEPS FEMORIS 2)

Arises Ant & lat shaft of femur to one hand's breadth above condyles

Inserts Quadriceps tendon to patella, via ligamentum patellae into tubercle of tibia

Action Extends knee

Nerve Post div of femoral N (L2,3,4)

VASTUS LATERALIS (QUADRICEPS FEMORIS 3)

Arises Upper intertrochanteric line, base of greater trochanter, lat linea aspera, lat supracondylar ridge & lat intermuscular septum

Inserts Lat quadriceps tendon to patella, via ligamentum patellae into tubercle of tibia

Action Extends knee

Nerve Post div of femoral N (L2,3,4)

VASTUS MEDIALIS (QUADRICEPS FEMORIS 4)

Arises Lower intertrochanteric line, spiral line, med linea aspera & med intermuscular septum

Inserts Med quadriceps tendon to patella & directly into med patella, via ligamentum patellae into tubercle of tibia

Action Extends knee. Stabilises patella

Nerve Post div of femoral N (L2,3,4)

ZYGOMATICUS MAJOR

Arises Ant surface of zygomatic bone

Inserts Modiolus at angle of mouth

Action Elevates & draws angle of mouth laterally

Nerve Buccal br of facial N (VII)

ZYGOMATICUS MINOR

Arises Lat infra-orbital margin

Inserts Skin & muscle of upper lip

Action Elevates & everts upper lip

Nerve Buccal br of facial N (VII)

***Aide memoir* for nerve supply of groups of muscles in head and neck**

All muscles of:	Supplied by:	Except:	Which is supplied by:
Pharynx	Pharyngeal plexus (IX, X, sympathetic)	Stylopharyngeus	Glossopharyngeal (IX)
Palate	Pharyngeal plexus (IX, X, sympathetic)	Tensor palati	Off N to med pterygoid (Vc)
Tongue	Hypoglossal (XII)	Palatoglossus	Pharyngeal plexus (IX, X, sympathetic)
Mastication	Mandibular (Vc)	Buccinator	Facial (VII)
Larynx	Recurrent laryngeal (X)	Cricothyroid	Ext br of superior laryngeal (X)
Facial expression & buccinator	Facial (VII)		

9: JOINTS

Classification of joints

Fibrous Fibrous tissue between bones
Primary cartilaginous Hyaline cartilage between bones
Secondary cartilaginous As for primary but fibrocartilage between the layers of hyaline cartilage (symphysis)
Synovial Joint cavity with synovial fluid. Hyaline cartilage on surface of bones. Articular disc can be present
Atypical synovial Joint cavity with synovial fluid. Fibrocartilage on surface of bones. Articular disc can be present

Types of synovial joint

Plane Sliding only
Hinge (ginglymus) One plane of movement
Modified hinge (bicondylar) One plane of movement + rotation
Condyloid (ellipsoid) Two planes of movement (circumduction)
Saddle condyloid (sella) Two planes of movement + controlled rotation
Pivot (trochoid) Rotation only. One plane of movement
Ball and socket (spheroidal) Multi-axial. Three planes of movement

Joints with interarticular fibrocartilaginous discs

Acromioclavicular (usually incomplete)
Femorotibial (knee) (incomplete—menisci)
Radiocarpal (wrist)
Sternoclavicular
Temporomandibular

Joints with double cavities separated by intra-articular ligaments—not fibrocartilaginous discs

Costovertebral (ribs 2–10)
Sternochondral (2nd rib)

Joints classified by type

Fibrous joints
Arytenocorniculate (can be synovial)
Costotransverse (ribs 11 and 12)
Cuboideonavicular (can be synovial)
Gomphosis (teeth)
Radio-ulnar (interosseous membrane)
Skull sutures
Tibiofibular (inferior)
Tibiofibular (interosseous membrane)

Primary cartilaginous joints
Costochondral
Sternochondral (1st rib)
Spheno-occipital

Secondary cartilaginous joints
Intervertebral
Manubriosternal
Sacrococcygeal
Symphysis pubis
Xiphisternal

Atypical synovial joints
Acromioclavicular
Sternochondral (ribs 2–7)
Sternoclavicular
Temporomandibular

Typical synovial
Acetabulofemoral (hip)
Atlanto-axial (dens & facets)
Atlanto-occipital
Calcaneocuboid
Carpometacarpal
Costotransverse (ribs 1–10)

Costovertebral
Crico-arytenoid
Crico-thyroid
Cuneocuboid
Cuneonavicular
Femorotibial (knee)
Glenohumeral (shoulder)
Humeroradial (elbow)
Humero-ulnar (elbow)
Intercarpal
Interchondral (cartilages 6–10)
Intercuneiform
Intermetacarpal
Intermetatarsal
Interphalangeal
Metacarpophalangeal
Metatarsophalangeal
Pisotriquetral
Radiocarpal (wrist)
Radio-ulnar (superior & inferior)
Talocalcaneal
Talocalcaneonavicular
Tarsometatarsal
Tibiofibular (superior)
Tibiotalal (ankle)
Zygapophyseal (intervertebral facet)

Unclassified

Intervertebral joints of Luschka

Named joints

All joints are paired except the following which are single midline joints:
Median atlanto-axial
Intervertebral
Manubriosternal
Spheno-occipital
Symphysis pubis
Xiphisternal

9

Fibrous
2 bones joined
by fibrous tissue

Examples: skull sutures, interosseous membranes, inferior tibiofibular joint

Cartilaginous
Primary
2 bones joined by
hyaline cartilage

Examples: costochondrial, first sternochondrial, spheno-occipital

Secondary
2 bone ends covered with
hyaline cartilage and a disc
of fibrocartilage between

All midline symphyses
Examples: intervertebral, pubis

Synovial
Atypical
Articular surface covered with
fibrocartilage
Examples: temporomandibular,
sternoclavicular,
2–7 sternochondrial

Typical
Articular surface covered with
hyaline cartilage
Examples: all synovial joints that are
not atypical

Types of movement in synovial joints

Plane
Examples: tarsus, carpus

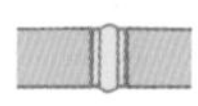

Hinge
Example: interphalangeal

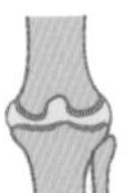

Modified hinge
Example: knee

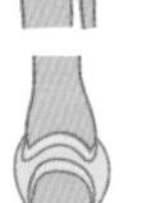

Condyloid
Example: metacarpophalangeal

Saddle condyloid
Example: first carpometacarpal

Ball and socket
Examples: hip, shoulder, sterno-clavicular, talocalcaneonavicular

Pivot
Examples: atlanto-axial, superior
and inferior radio-ulnar

Classification of joints

ACETABULOFEMORAL (hip)
Classification Synovial
Type Ball & socket
Articulation Acetabulum with femur

ACROMIOCLAVICULAR
Classification Atypical synovial
Type Plane
Articulation Acromion with clavicle
Notes Often an articular disc in upper part of jnt, usually incomplete

ANKLE (see tibiotalal)

ARYTENOCORNICULATE (larynx)
Classification Fibrous or synovial
Articulation Arytenoid cartilage with corniculate cartilage

ATLANTO-AXIAL—LATERAL
Classification Synovial
Type Plane
Articulation Articular facets of atlas with axis

ATLANTO-AXIAL—MEDIAN
Classification Synovial
Type Pivot
Articulation Dens of axis with atlas
Notes Second cavity (bursa) posteriorly

ATLANTO-OCCIPITAL
Classification Synovial
Type Condyloid
Articulation Atlas with occipital bone

CALCANEOCUBOID (midtarsal)
Classification Synovial
Type Saddle condyloid
Articulation Calcaneus with cuboid
Notes This is one part of midtarsal jnt. Other is talonavicular part of talocalcaneonavicular

CARPOMETACARPAL—FINGERS 2–5 (including intermetacarpal)
Classification Synovial
Type Plane
Articulation Carpal bones with MCs & between MCs
Notes Usually continuous cavity between CMC, intermetacarpal & intercarpal jnts

CARPOMETACARPAL—THUMB
Classification Synovial
Type Saddle condyloid
Articulation Trapezium with 1st MC
Notes Joint is separate from others in hand

COSTOCHONDRAL
Classification Primary cartilaginous
Articulation Bony rib with costal cartilage

COSTOTRANSVERSE—RIBS 1–10
Classification Synovial
Type Plane
Articulation Med facet of tubercle of rib with transverse process of own vertebra

COSTOTRANSVERSE—RIBS 11, 12
Classification Fibrous (ligamentous)
Articulation Tubercle of rib with transverse process of own vertebra

COSTOVERTEBRAL
Classification Synovial
Type Plane
Articulation Head of rib with vertebral body
Notes 1st rib articulates with T1 vertebra only (single cavity jnt). Ribs 2–10 with own vertebra & one above (double cavity jnts separated by intra-articular lig). Ribs 11 & 12 with own vertebra only (single cavity jnts)

CRICO-ARYTENOID (larynx)
Classification Synovial
Type Features of ball & socket
Articulation Cricoid cartilage with arytenoid cartilage

CRICOTHYROID (larynx)
Classification Synovial
Type Plane (rotational). Two together give hinge movement
Articulation Facet on side of cricoid cartilage with inf horn of thyroid cartilage

ELBOW (see humeroradial & humero-ulnar)

FEMOROTIBIAL (knee)
Classification Synovial
Type Modified hinge
Articulation Femur with tibia
Notes Menisci are incomplete discs of fibrocartilage

GLENOHUMERAL (shoulder)
Classification Synovial
Type Ball & socket
Articulation Glenoid fossa of scapula with humerus

GOMPHOSIS (dento-alveolar)
Classification Fibrous
Articulation Tooth with bone of jaw

HIP (see acetabulofemoral)

HUMERORADIAL (elbow)
Classification Synovial
Type Hinge
Articulation Capitulum of humerus with radial head
Notes Jnt cavity is shared with humero-ulnar & sup radio-ulnar jnts

HUMERO-ULNAR (elbow)
Classification Synovial
Type Hinge
Articulation Trochlea of humerus with ulnar trochlear notch
Notes Jnt cavity is shared with humeroradial & sup radio-ulnar jnts

INTERCARPAL (MIDCARPAL)
(see also pisotriquetral)
Classification Synovial
Type Plane individually but together give effective mixture of condyloid, saddle condyloid & ball & socket
Articulation Between scaphoid, lunate, triquetral, hamate, capitate, trapezoid & trapezium
Notes Single cavity between the seven bones usually communicating also with CMC & intercarpal jnts of fingers 2–5

INTERCHONDRAL
Classification Synovial
Type Plane
Articulation Between costal cartilages 6/7, 7/8, 8/9, 9/10

INTERMETATARSAL
Classification Synovial
Type Plane
Articulation Between MTs

INTERPHALANGEAL (fingers & toes)
Classification Synovial
Type Hinge
Articulation Between phalanges

INTERTARSAL–CUBOIDEONAVICULAR
Classification Fibrous (can be synovial)
Articulation Cuboid with navicular

INTERTARSAL–CUNEOCUBOID
Classification Synovial
Type Plane
Articulation Lat cuneiform with cuboid
Notes Cuneocuboid shares cavity with cuneonavicular & intercuneiform jnts

INTERTARSAL—CUNEONAVICULAR
Classification Synovial
Type Plane
Articulation Cuneiforms with navicular

9

Notes Cuneonavicular shares cavity with cuneocuboid and intercuneiform jnts

INTERTARSAL—INTERCUNEIFORM
Classification Synovial
Type Plane
Articulation Between cuneiforms
Notes Intercuneiform jnts share cavity with cuneonavicular & cuneocuboid jnts

INTERVERTEBRAL
Classification Secondary cartilaginous
Articulation Between vertebral bodies

INTERVERTEBRAL JOINTS OF LUSCHKA (neurocentral or uncovertebral)
Classification Unclassified
Type Unclassified
Articulation Lateral lip of upper surface of C3–7 & T1 vertebrae with adjacent vertebral body above
Notes Often small cavity which is degenerative (not synovial)

KNEE (see femorotibial)

MANUBRIOSTERNAL
Classification Secondary cartilaginous
Articulation Manubrium with sternum
Notes May cavitate to give appearance of synovial jnt

METACARPOPHALANGEAL
Classification Synovial
Type Condyloid
Articulation MCs with phalanges

METATARSOPHALANGEAL
Classification Synovial
Type Condyloid
Articulation MTs with phalanges

MIDTARSAL (see both talonavicular part of talocalcaneonavicular and calcaneocuboid)

PISOTRIQUETRAL
Classification Synovial
Type Plane
Articulation Pisiform with triquetral

RADIOCARPAL (wrist)
Classification Synovial
Type Condyloid
Articulation Radius & triangular fibrocartilaginous articular disc with scaphoid, lunate & triquetral

RADIO-ULNAR—INFERIOR
Classification Synovial
Type Pivot
Articulation Radius with ulna
Notes Cavity separated from cavity of wrist by triangular fibrocartilaginous disc

RADIO-ULNAR—INTEROSSEOUS MEMBRANE & OBLIQUE CORD
Classification Fibrous
Articulation Radius with ulna

RADIO-ULNAR—SUPERIOR
Classification Synovial
Type Pivot
Articulation Radius with ulna
Notes Cavity is continuous with humero-ulnar & humeroradial jnts

SACROCOCCYGEAL
Classification Secondary cartilaginous
Articulation Sacrum with coccyx

SACRO-ILIAC
Classification Synovial
Type Plane
Articulation Sacrum with iliac bone

SHOULDER (see glenohumeral)

SKULL SUTURES
Classification Fibrous
Articulation Between diploae of skull

SPHENO-OCCIPITAL
Classification Primary cartilaginous
Articulation Sphenoid with basi-occiput

STERNOCHONDRAL (sternocostal)
Classification 1st: primary cartilaginous; 2nd–7th: atypical synovial
Type 2nd–7th: plane
Articulation 1st rib with manubrium. 3rd–7th ribs with sternum. 2nd with both
Notes 2nd jnt has two cavities separated by intra-articular lig

STERNOCLAVICULAR (manubrioclavicular)
Classification Atypical synovial
Type Features of ball & socket
Articulation Clavicle with manubrium
Notes Separated into two cavities by fibrocartilaginous disc

SUBTALAR (see both talocalcanean & talocalcaneal part of talocalcaneonavicular)

SYMPHYSIS PUBIS
Classification Secondary cartilaginous
Articulation Between pubic bones
Notes May cavitate

TALOCALCANEAN (subtalar)
Classification Synovial
Type Plane (effectively ball & socket)
Articulation Talus with calcaneus
Notes This is a posterior, separate jnt that is one-half of subtalar jnt

TALOCALCANEONAVICULAR
Classification Synovial
Type Features of ball & socket
Articulation Talus with calcaneus & navicular
Notes This is a two part jnt. Talocalcaneal part (two facets) is part of subtalar jnt, other being talocalcanean. Talonavicular part (one facet) is part of midtarsal jnt, other being calcaneocuboid.

TARSOMETATARSAL
Classification Synovial
Type Plane
Articulation Tarsal bones with MTs

TEMPOROMANDIBULAR
Classification Atypical synovial
Type Condyloid
Articulation Temporal bone with mandible
Notes Separated into two cavities by fibrocartilaginous disc

TIBIOFIBULAR—INFERIOR
Classification Fibrous
Articulation Tibia with fibula

TIBIOFIBULAR—INTEROSSEOUS MEMBRANE
Classification Fibrous
Articulation Tibia with fibula

TIBIOFIBULAR—SUPERIOR
Classification Synovial
Type Plane
Articulation Tibia with fibula

TIBIOTALAL (ankle)
Classification Synovial
Type Hinge
Articulation Tibia with talus

WRIST (see radiocarpal & interarticular disc of inferior radio-ulnar)

XIPHISTERNAL
Classification Secondary cartilaginous
Articulation Xiphoid with sternum

ZYGAPOPHYSEAL (intervertebral facet)
Classification Synovial
Type Plane
Articulation Between intervertebral facets

10: OSSIFICATION TIMES

Ossification times

Bones (number if unpaired)	Forms in membrane (M) or cartilage (C)	Centres primary (P) or secondary (S)	Site	Centre appears at: Gestation (weeks/months)	After birth	Fused by
Mandible (1)	M	P	Near mental foramen (each side)	6 W		Symphysis menti 1–3 Y
Hyoid (1)	C	P	Greater cornu (each side)	8–9 M		
	C	S	Body (2 centres)	9 M		
	C	S	Lesser cornu (each side)		Puberty	
Occiput (1)	M	P	Squamous (each side)	8 W		
	C	P	Lateral (each side)	8 W		
	C	P	Basilar	8 W		
Sphenoid (1)	M + C	P	Approximately 14 centres	8 W–4 M		
Temporal	M	P	Squamous	8 W		
	M	P	Tympanic	3 M		
	C	P	Petromastoid (several centres)	5 M		
Parietal	M	P	Near tuberosity (2 centres)	7 W		
Frontal (2 → 1)	M	P	Near each tuberosity (2 centres, one each side)	8 W		Metopic suture 2 Y
Ethmoid (1)	C	P	Labyrinth (one each side)	5 M		
	C	P	Perpendicular plate/crista galli		1 Y	
Inf concha	M	P		5 M		
Lacrimal	M	P		4 M		
Nasal	M	P		3 M		
Vomer (1)	M	P	(2 centres)	8 W		
Maxilla	M	P	(3 centres)	6–8 W		
Palatine	M	P	Perpendicular plate	8 W		

Zygomatic	M	P		8 W		
Ear ossicles	C	P	Stapes	4 M		
	C	P	Malleus	4 M		
	C	P	Incus	4 M		
Scapula	C	P	Body	8 W		
	C	S	Coracoid process		1 Y	15 Y
	C	S	Subcoracoid		Puberty	20 Y
	C	S	Medial border		Puberty	20 Y
	C	S	Glenoid (lower rim)		Puberty	20 Y
	C	S	Acromion (2 centres)		Puberty	20 Y
	C	S	Inferior angle		Puberty	20 Y
Clavicle	M	P	Medial & lateral (2 centres)	5 W		
	M	S	Sternal end		Late teens	20 Y
Humerus (upper end is growing end)	C	P	Shaft	8 W		
	C	S	Head		6 M	Upper epiphysis 18–20 Y
	C	S	Greater tuberosity		2 Y	
	C	S	Lesser tuberosity		5 Y	
	C	S	Capitulum & lat ridge of trochlea		1 Y	
	C	S	Medial trochlea		10 Y	
	C	S	Medial epicondyle		5 Y	Lower epiphysis 14–16 Y
	C	S	Lateral epicondyle		12 Y	
Radius (lower end is growing end)	C	P	Shaft	8 W		
	C	S	Head		4 Y	14–17 Y
	C	S	Distal end		1 Y	17–19 Y
Ulna (lower end is growing end)	C	P	Shaft	8 W		
	C	S	Olecranon (2 centres)		9 Y	14–16 Y
	C	S	Distal end		5 Y	17–18 Y
Carpus	C	P	Capitate		2 M	
	C	P	Hamate		3 M	
	C	P	Triquetral		3 Y	
	C	P	Lunate		4 Y	

continued on p. 188

Ossification times *continued*

Bones (number if unpaired)	Forms in membrane (M) or cartilage (C)	Centres primary (P) or secondary (S)	Site	Centre appears at: Gestation (weeks/months)	After birth	Fused by
	C	P	Scaphoid		4–5 Y	
	C	P	Trapezium		4–5 Y	
	C	P	Trapezoid		4–5 Y	
	C	P	Pisiform		9–12 Y	
Metacarpal (1st)	C	P	Shaft	9 W		
	C	S	Base		3 Y	15–17 Y
Metacarpals (2nd–5th)	C	P	Shaft	9 W		
	C	S	Head		2 Y	15–19 Y
Phalanges (hand)	C	P	Shaft	8–12 W		
	C	S	Base		2–4 Y	15–18 Y
Innominate	C	P	Pubis (superior ramus)	4 M		7–8 Y
	C	P	Ischium (body)	4 M		7–8 Y
	C	P	Ilium (above greater sciatic notch)	2 M		7–8 Y
	C	S	Iliac crest (2 centres)		Puberty	15–25 Y
	C	S	Acetabulum (2 centres)		Puberty	15–25 Y
	C	S	Anterior superior iliac spine		Puberty	15–25 Y
	C	S	Ischial tuberosity		Puberty	15–25 Y
	C	S	Pubic symphysis		Puberty	15–25 Y
Femur (lower end is growing end)	C	P	Shaft	7 W		
	C	S	Greater trochanter		4 Y	
	C	S	Lesser trochanter		12 Y	
	C	S	Head		6 M	14–17 Y
	C	S	Distal end	9 M		16–18 Y
Patella	C	P	(Several centres)		3–6 Y	Puberty
	C	S	Superolaterally		6 Y	Puberty

Tibia (upper end is growing end)	C	P	Shaft	7 W		
	C	S	Plateau	9 M		16–18 Y
	C	S	Distal end		1 Y	15–17 Y
	C	S	Tuberosity		12 Y	13–14 Y
Fibula (upper end is growing end)	C	P	Shaft	8 W		
	C	S	Distal end		1 Y	15–17 Y
	C	S	Head		3–4 Y	17–19 Y
Talus	C	P		6 M		
Calcaneus	C	P		3 M		
	C	S			6–8 Y	14–16 Y
Navicular	C	P			3 Y	
Cuneiform lateral	C	P			1 Y	
Cuneiform medial	C	P	(May have 2 centres)		2 Y	
Cuneiform intermediate	C	P			3 Y	
Cuboid	C	P		9 M		
Metatarsal (1st)	C	P	Shaft	9 W		17–20 Y
	C	S	Base		3 Y	17–20 Y
Metatarsals (2nd–5th)	C	P	Shaft	9 W		17–20 Y
	C	S	Head		3–4 Y	17–20 Y
Phalanges (foot)	C	P	Shaft	9–15 W		18 Y
	C	S	Base		2–8 Y	18 Y

continued on p. 190

Closure of skull sutures

Ant fontinelle: closes 18 M; post fontinelle: closes 6 M–1 Y

Notes: (1) All bones are paired unless otherwise stated. (2) Single centre of ossification unless specified otherwise. (3) Variability of ossification usually a sex difference, females appearing and uniting earlier. (4) Fusion times for epiphyses are given if clinically relevant

Ossification times *continued*—eruption of teeth

Eruption of teeth	Incisor		Canine	Premolar	Molar
	(upper)	(lower)			
First dentition (months)	7, 8	6, 9	18		12, 24
Second dentition* (years)	7, 8	7, 8	11	9, 10	6, 12, 18

* Lower teeth erupt slightly earlier.

11: FORAMINA—SKULL AND SPINE

Notes: (1) Most smaller emissary veins and meningeal arterial supplies have been omitted. (2) All structures are paired unless otherwise indicated.

AQUEDUCT OF THE VESTIBULE
Site In post aspect of petrous temporal bone in post cranial fossa, 1 cm post to int acoustic meatus
Contains Endolymphatic duct & sac, small art & V

CAROTID CANAL
Site In inf surface of petrous temporal bone in middle cranial fossa
Contains Int carotid art enters with sympathetic plexus on it. Int carotid venous plexus connecting cavernous sinus & int jugular vein

CONDYLAR CANAL
Site In lower sigmoid groove in occipital bone in post cranial fossa. Exits at condylar fossa behind condyle (not always present)
Contains Emissary V connecting sigmoid sinus & occipital Vs. Meningeal br of occipital art

CRIBRIFORM FORAMINA
Site In cribriform plate of ethmoid bone in ant cranial fossa
Contains Olfactory filaments & ant ethmoidal N & vessels

EXTERNAL ACOUSTIC MEATUS
Site Inner two thirds osseous within tympanic part of temporal bone and limited medially by tympanic membrane. Outer one third cartilaginous in continuity with auricular cartilage.
Contains Nil (Ceruminous glands and hair follicles mostly in the skin of the cartilaginous part).

FACIAL CANAL
Site In petrous temporal bone leading from int acoustic meatus to stylomastoid foramen
Contains Facial N (VII)

FORAMEN CAECUM (unpaired)
Site Between frontal crest of frontal bone & crista galli of ethmoid bone in ant cranial fossa
Contains Emissary Vs connecting nose & sup sagittal sinus

FORAMEN LACERUM
Site Between sphenoid, apex of petrous temporal & basilar occipital bones in middle cranial fossa
Contains Int carotid art enters behind & exits above. Greater petrosal N enters behind/above & leaves ant as N of the pterygoid canal

FORAMEN MAGNUM (unpaired)
Site In occipital bone in post cranial fossa
Contains Medulla oblongata, meninges, vertebral arts, ant & post spinal arts, spinal roots of accessory Ns (XI), sympathetic plexus on vertebral art, apical ligament of dens, tectorial membrane

FORAMEN OVALE
Site In greater wing of sphenoid bone in middle cranial fossa
Contains Mandibular N (Vc), lesser petrosal N, accessory meningeal art

FORAMEN ROTUNDUM
Site In greater wing of sphenoid bone in middle cranial fossa
Contains Maxillary N (Vb)

FORAMEN SPINOSUM
Site In greater wing of sphenoid bone in middle cranial fossa
Contains Middle meningeal vessels, meningeal br of mandibular N (Vc)

FORAMEN TRANSVERSARIUM

Site In pedicle of cervical vertebrae bordered by—lat: intertubercular lamella (costotransverse bar), med: body of vertebra

Contains Vertebral art & V in C1–6. Vein only in C7

GREATER PALATINE FORAMEN

Site Between maxilla & palatine bone at lat edge of hard palate

Contains Greater palatine N & vessels

GREATER PETROSAL HIATUS

Site On anterior surface of petrous temporal bone and extending as a small groove on that bone. It lies just medial to the lesser petrosal hiatus

Contains Greater petrosal nerve

HYPOGLOSSAL CANAL

Site In occipital bone above condyle in post cranial fossa

Contains Hypoglossal N (XII) & meningeal br of ascending pharyngeal art

INCISIVE CANAL

Site In ant maxilla extending from nose to incisive foramina

Contains Nasopalatine N, greater palatine vessels

INCISIVE FORAMEN

Site Midline, in ant hard palate. Openings of incisive canals into incisive fossa

Contains Nasopalatine N, greater palatine vessels

INCISIVE FOSSA (unpaired)

Site Median, in ant hard palate leading upwards to incisive foramina

Contains Nasopalatine Ns, greater palatine vessels

INFERIOR ORBITAL FISSURE

Site Between greater wing of sphenoid bone & maxilla

Contains Infra-orbital & zygomatic brs of maxillary N (Vb), infra-orbital vessels, inf ophthalmic Vs, orbital brs of pterygopalatine ganglion

INFRA-ORBITAL CANAL

Site Within orbital aspect of maxilla

Contains Infra-orbital N & vessels

INFRA-ORBITAL FORAMEN

Site Below infra-orbital margin in maxilla. Ant opening of infra-orbital canal

Contains Infra-orbital N & vessels

INTERNAL ACOUSTIC MEATUS

Site In post surface of petrous temporal bone in post cranial fossa

Contains Facial N (VII), nervus intermedius, vestibulocochlear N (VIII), labyrinthine art

INTERVERTEBRAL FORAMEN

Site Between vertebrae, bordered by—sup & inf: pedicles of vertebrae, ant: vertebral bodies & intervertebral disc, post: lig flavum covering sup & inf articular processes

Contains Spinal art & V, dorsal root ganglion, spinal N. Nerves C1–7 emerge via foramen above same numbered vertebra; nerve C8 exits below C7 vertebra & below this all nerves emerge via foramen below the same numbered vertebra

JUGULAR FORAMEN

Site Between jugular fossa of petrous temporal bone & occipital bone in post cranial fossa

Contains Glossopharyngeal N (IX), vagus (X), accessory N (XI), inf petrosal & sigmoid sinuses enters it, int jugular V emerges below

LESSER PALATINE FORAMINA
Site Two or three foramina in med & inf aspects of pyramidal process of palatine bone
Contains Lesser palatine Ns & vessels

LESSER PETROSAL HIATUS
Site Small hole on anterior surface of petrous temporal bone lying just lateral to the greater petrosal hiatus
Contains Lesser petrosal nerve

MANDIBULAR CANAL (INFERIOR ALVEOLAR CANAL)
Site In body & ramus of mandible between mandibular & mental foramina
Contains Inf alveolar N & vessels

MANDIBULAR FORAMEN (INFERIOR ALVEOLAR FORAMEN)
Site Med aspect of ramus of mandible, overlapped anteromedially by lingula
Contains Inf alveolar N & vessels

MASTOID FORAMEN
Site In petrous temporal bone in post cranial fossa, post to sigmoid groove. Exits behind mastoid process
Contains Emissary V connecting sigmoid sinus & occipital Vs, meningeal br of occipital art

MENTAL FORAMEN
Site Outer aspect of ant ramus of mandible by second premolar tooth, leading from mandibular (inf alveolar) canal
Contains Mental N & vessels

NASOLACRIMAL CANAL
Site Between lacrimal bone & maxilla at ant/inf/med corner of orbit
Contains Nasolacrimal duct

OPTIC CANAL
Site In body of sphenoid bone in middle cranial fossa between body & two roots of lesser wing
Contains Optic N (II), dural sheath, ophthalmic art

PALATOVAGINAL CANAL
Site Between upper surface of sphenoidal process of palatine bone & lower surface of vaginal process of root of med pterygoid plate in base of skull
Contains Pharyngeal Ns from maxillary (Vb) and pterygopalatine ganglion & pharyngeal br of maxillary art

PETROSQUAMOUS FISSURE
Site Between squamous temporal bone & tegmen tympani (petrous temporal bone)
Contains No structures

PETROTYMPANIC FISSURE
Site Between tympanic part (plate) of temporal bone & tegmen tympani (also part of temporal bone) in base of skull
Contains Chorda tympani, ant lig of malleus, ant tympanic br of maxillary art

PTERYGOID CANAL
Site In pterygoid process of sphenoid bone connecting ant wall of foramen lacerum to pterygopalatine fossa
Contains N & art of pterygoid canal

PTERYGOMAXILLARY FISSURE
Site Between lat pterygoid plate & post surface of maxilla connecting infra-temporal & pterygopalatine fossae, continuous above with post end of inf orbital fissure
Contains Terminal brs of maxillary art, passing in, post sup alveolar Ns passing out

SPHENOIDAL FORAMEN
Site In greater wing of sphenoid in middle cranial fossa med to foramen ovale (40% of skulls) (venous foramen of Vesalius)
Contains Emissary V connecting cavernous sinus & pterygoid plexus

SPHENOPALATINE FORAMEN
Site Between body of sphenoid bone & sphenopalatine notch of palatine bone (sup border of perpendicular plate & orbital & sphenoidal processes). In med wall of pterygopalatine fossa
Contains Sphenopalatine art, nasopalatine & sup nasal Ns from pterygopalatine fossa

SQUAMOTYMPANIC FISSURE
Site Between tympanic part (plate) of temporal bone & mandibular fossa (squamous temporal bone) in base of skull. It is divided by tegmen tympani (petrous temporal bone) into petrotympanic and petrosquamous fissures
Contains Deep auricular br of maxillary artery

STYLOMASTOID FORAMEN
Site Between styloid & mastoid processes of temporal bone in base of skull
Contains Facial N (VII) & stylomastoid br of post auricular art

SUPERIOR ORBITAL FISSURE
Site Between body & lesser & greater wings of sphenoid bone in middle cranial fossa
Contains Ophthalmic N (Va) (lacrimal, frontal, nasociliary brs), ophthalmic Vs, oculomotor N (sup & inf divs) (III), trochlear N (IV), abducent N (VI), sympathetic fibres, brs of middle meningeal & lacrimal arts

SUPRA-ORBITAL FORAMEN
Site In supra-orbital margin of frontal bone, 2cm from midline
Contains Supra-orbital N & vessels

VERTEBRAL FORAMEN (unpaired)
Site Bordered by—ant: body of vertebra, post: laminae, lat: pedicles & articular processes. Collectively making the spinal canal
Contains Spinal cord/cauda equina, dura, archnoid & pia mater, cerebrospinal fluid, internal vertebral venous plexus & spinal arts

VOMEROVAGINAL CANAL
Site Between lower aspect of ala of vomer & upper aspect of vaginal process of root of med pterygoid plate in base of skull (not always present)
Contains Pharyngeal br of sphenopalatine art

ZYGOMATICOFACIAL FORAMEN
Site In lat surface of zygomatic bone
Contains Zygomaticofacial N & vessels

ZYGOMATICO-ORBITAL FORAMEN
Site In orbital surface of zygomatic bone
Contains Zygomatic br of maxillary N (Vb)

ZYGOMATICOTEMPORAL FORAMEN
Site In posteromedial surface of zygomatic bone
Contains Zygomaticotemporal N & vessels

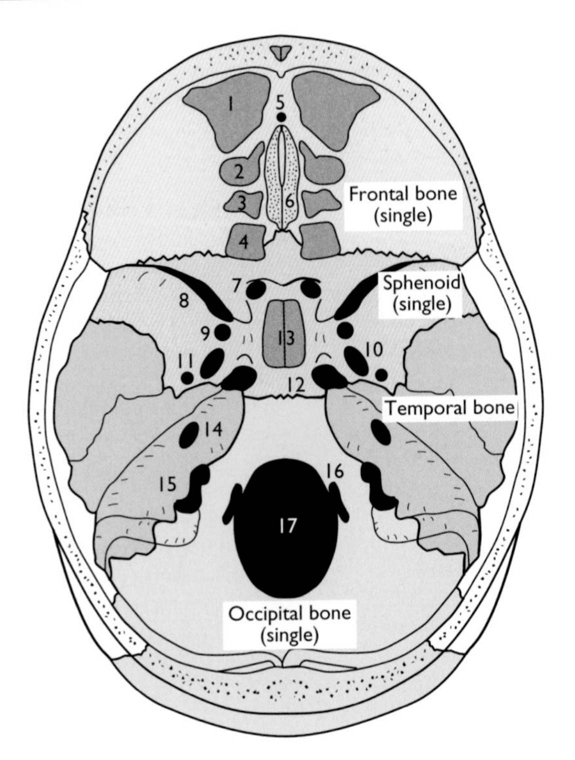

Foramina and air sinuses

1 Frontal air sinuses
2 Anterior ethmoidal air sinuses
3 Middle ethmoidal air sinuses
4 Posterior ethmoidal air sinuses
5 Foramen caecum (single midline)
6 Cribriform plate of ethmoid
7 Optic canal
8 Superior orbital fissure
9 Foramen rotundum
10 Foramen ovale
11 Foramen spinosum
12 Foramen lacerum
13 Sphenoid air sinuses
14 Internal acoustic (auditory) meatus
15 Jugular foramen
16 Hypoglossal canal
17 Foramen magnum (single midline)

Internal base of skull

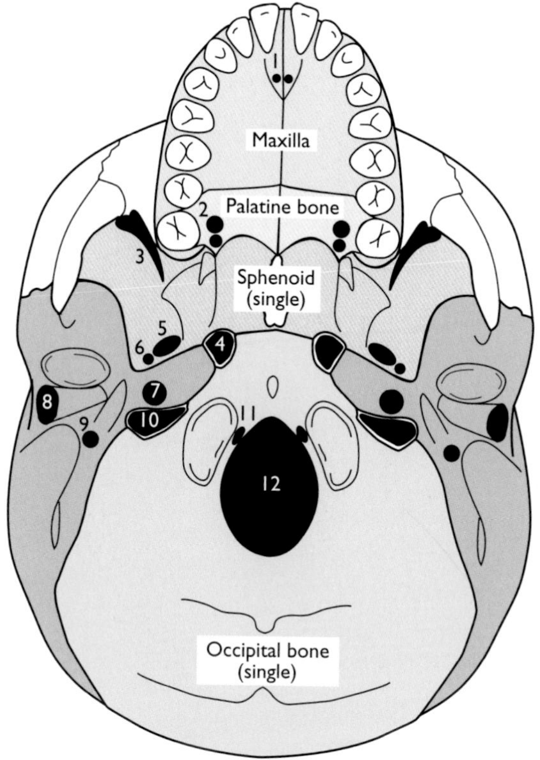

Foramina

1 Left and right incisive foramina in single midline incisive fossa
2 Greater and lesser palatine foramina
3 Inferior orbital fissure
4 Foramen lacerum
5 Foramen ovale
6 Foramen spinosum
7 Carotid canal
8 External acoustic (auditory) meatus
9 Stylomastoid foramen
10 Jugular foramen
11 Hypoglossal canal
12 Foramen magnum (single midline)

External base of skull

NOTES

12: POSITION OF STRUCTURES ACCORDING TO VERTEBRAL LEVELS

Notes: (1) A /(oblique stroke) between two levels indicates that the structure lies at a level between these two vertebrae. (2) A – (dash) between two levels indicates that the structure occupies the equivalent level to these vertebrae inclusively.

C1 Spinal root of accessory nerve crosses transverse process of atlas
Open mouth and dens

C2 Superior cervical ganglion

C3 Body of hyoid bone

C4 Upper border of thyroid cartilage
Bifurcation of common carotid arteries

C6 Cricoid cartilage
Larynx becomes trachea
Pharynx becomes oesophagus
Middle cervical ganglion
Vertebral artery enters foramen transversarium of C6 vertebra
Carotid tubercle of Chassaignac
Inferior thyroid artery crosses to thyroid gland

C7 First clearly palpable spinous process (vertebra prominens)
Stellate/inferior cervical ganglion

T2 Superior border of scapula

T2/3 Suprasternal notch

T3 Medial end of spine of scapula
End of oblique fissure of lung posteriorly at spine of T3

T3/4 Top of arch of aorta

T3–4 Manubrium sterni

T4/5 Manubriosternal angle of Louis
Anterior aspect of second rib
Under surface of arch of aorta
Ligamentum arteriosum
Left recurrent laryngeal nerve
Bifurcation of trachea
Division of pulmonary trunk
Azygos vein (arch) enters superior vena cava

T5 Thoracic duct crosses midline

T5–8 Sternum

T6 Upper border of liver

T7 Inferior angle of scapula
Accessory hemiazygos vein crosses midline to azygos vein

T8 Caval opening in diaphragm
- Inferior vena cava
- Right phrenic nerve

Left phrenic nerve pierces diaphragm lat to central tendon
Hemiazygos vein crosses to right to join azygos vein

T8/9 Sternoxiphisternal joint

T9 Superior epigastic vessels traverse diaphragm
Xiphoid

T10 Oesophageal opening in diaphragm
- Oesophagus
- Brs of left gastric vessels
- Anterior and posterior vagi

T12 Aortic 'opening' in diaphragm posterior to median arcuate lig
- Aorta
- Azygos & hemiazygos veins
- Thoracic duct

	Origin of coeliac axis (lower border of T12) Splanchnic nerves pierce crura of diaphragm Sympathetic trunk passes posterior to medial arcuate ligament Subcostal neurovascular bundle passes posterior to lateral arcuate lig
L1	Transpyloric plane (of Addison) (half way between suprasternal notch and symphysis pubis) • L1 vertebral body • End of spinal cord (lower border of L1) • Tip of 9th costal cartilage • Fundus of gallbladder • Pylorus of stomach • Second part of duodenum • Sphincter of Oddi • Neck of pancreas • Origin of portal vein • Hila of kidneys • Origin of superior mesenteric artery • Attachment of transverse mesocolon
L1/2	Origin of renal arteries
L2	Subcostal plane • Formation of azygos and hemiazygos veins • Duodenojejunal flexure, ligament of Treitz (upper border of L2) • Origin of gonadal arteries
L3	Origin of inferior mesenteric artery
L3/4	Umbilicus
L4	Supracristal plane (iliac crests) • Bifurcation of aorta
L5	Formation of inferior vena cava
S2	Sacral dimple Mid point of sacro-iliac joint Post superior iliac spine Dural sac ends Inferior attachment of small bowel mesentery
S3	Start of rectum
S4	Sacral hiatus End of vertebral canal
Co1	Filum terminale inserts

13: PHARYNGEAL DERIVATIVES

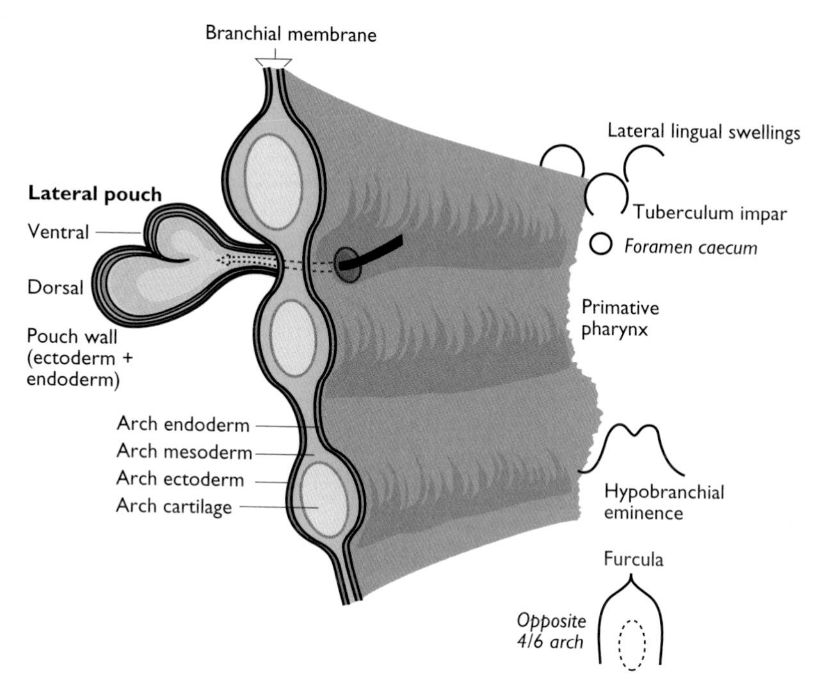

Pharyngeal derivatives

Pharyngeal derivatives

	Arch derivatives			Lateral pouch derivatives		Artery	Nerve
	Mesoderm		Endoderm	Endoderm	Ectoderm		
Arch	Cartilages, bones, ligaments	Muscles					
1 'Mandibular' arch cartilage = Meckel's	Incus Malleus Anterior lig of malleus Sphenomandibular lig (Lingula) (Mandible*)	Masseter Temporalis Pterygoids Mylohyoid Anterior belly of digastric Tensor veli palatini Tensor tympani	Mucous membrane & glands of anterior 2/3 of tongue	Auditory tube Inner layer of tympanic membrane (Part of middle ear) (Mastoid antrum)	External acoustic meatus Outer layer of tympanic membrane (Tragus of ear) (Skin of lower face)	Part of maxillary artery	Mandibular division of trigeminal (Vc)
2 'Hyoid' arch cartilage = Reichert's	Upper body & lesser cornu of hyoid Stylohyoid ligament Styloid process Stapes	Stapedius Stylohyoid Posterior belly of digastric Muscles of facial expression including buccinator & platysma	—	Supratonsillar fossa Tonsillar crypts Surface epithelium of tonsil** (Part of middle ear)	Overgrowth of ectoderm over arches 3, 4 & 6	Stapedial artery	Facial (VII)

3 'Thyrohyoid' arch	Inferior body & greater cornu of hyoid	Stylopharyngeus	Mucous membrane & glands of posterior 1/3 of tongue	Ventral: epithelial cells of thymus** Dorsal: inferior parathyroid	—	Internal carotid artery (including carotid sinus)	Glossopharyngeal (IX)
4	Thyroid cartilage	Palatoglossus Palatopharyngeus Salpingopharyngeus Cricothyroid Levator veli palatini Striated of oesophagus Pharyngeal constrictors	Valleculae & anterior epiglottis	Ventral: ultimo-branchial bodies† Dorsal: superior parathyroid	—	Right: part of right subclavian artery Left: aortic arch	Vagus (X) Pharyngeal & superior laryngeal branches
6	Cricoid cartilage Vocal ligs Arytenoid, corniculate & cuneiform cartilages	Cricopharyngeus All intrinsic muscles of larynx	—	Lung buds	—	Ventral: pulmonary artery Dorsal: ductus arteriosus	Vagus (X) Recurrent laryngeal branch

* The mandible forms in membrane around the ventral aspect of the first arch cartilage.

** The lymphoid tissue of the tonsil and thymus arises from the surrounding mesenchyme and is not arch derivative.

† Ultimobranchial bodies develop from ventral parts of fourth (and possibly fifth) pouch and fuse with the developing thyroid to give parafollicular (C) cells which produce calcitonin.

Notes: (1) The thyroid gland arises from between the first and second arch as a diverticulum (thyroglossal duct) which grows downwards leaving the foramen caecum at its origin. (2) The epiglottis comes from the inferior part of the hypobranchial eminence and is thus not a true arch derivative. (3) Bracketed information is of additional interest

13

14: SURFACE ANATOMY AND KEY AREAS

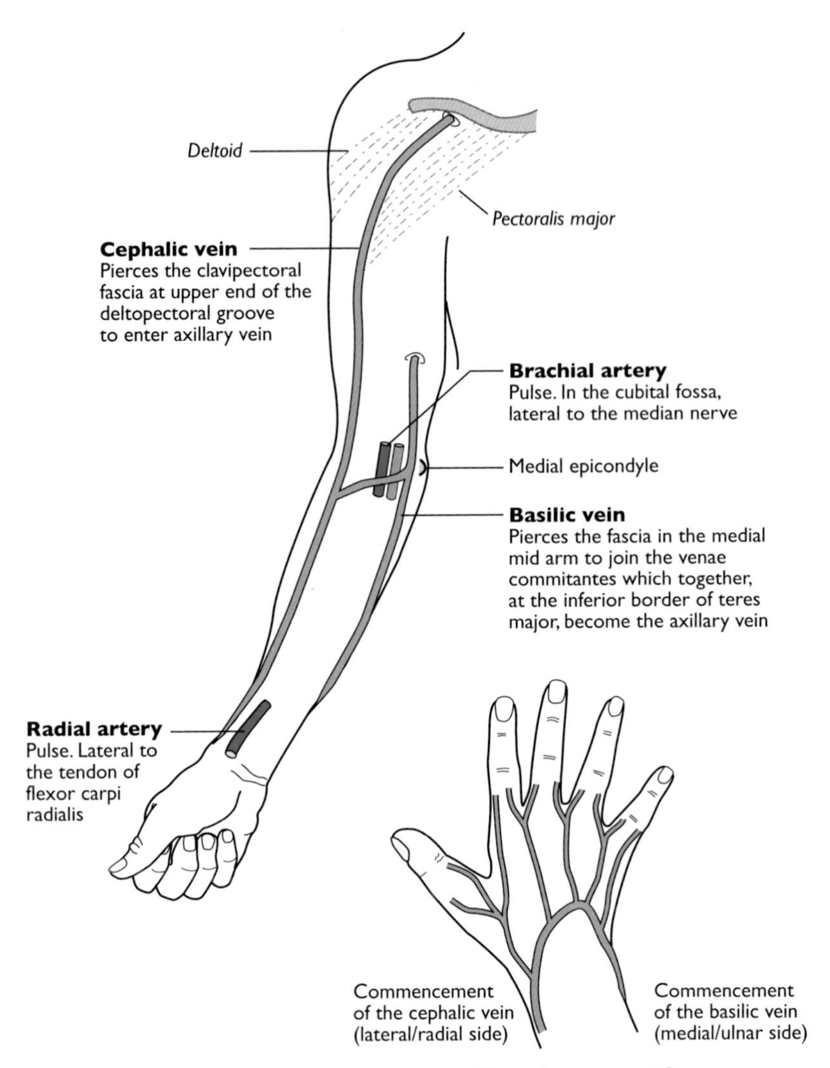

Superficial veins and pulses in upper limb

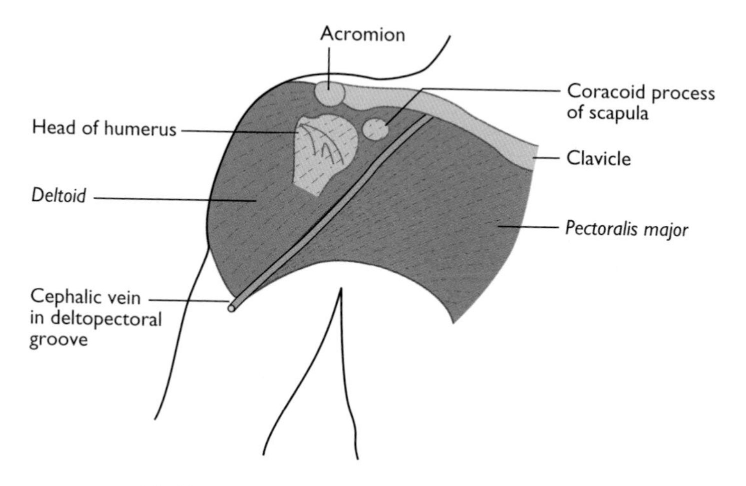

Palpable around the shoulder are:
The acromion
The head of the humerus
The coracoid process
The clavicle

Landmarks around shoulder

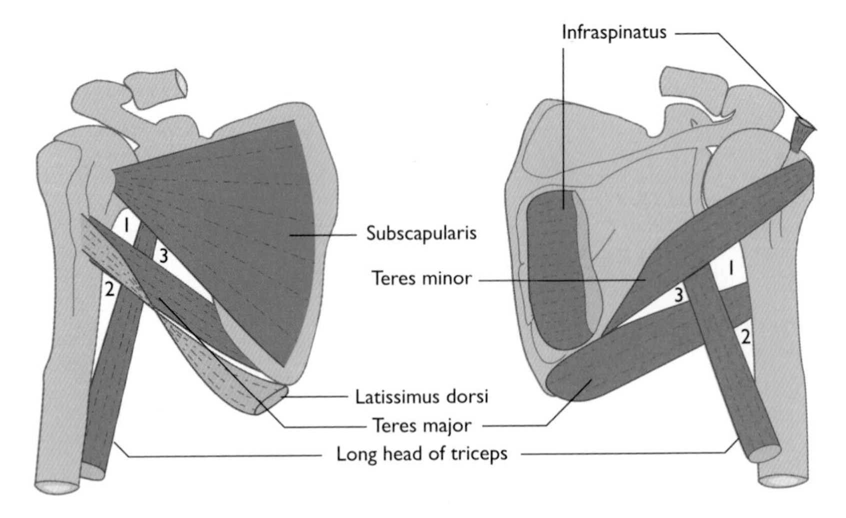

Anterior view
Right upper limb

Posterior view
Right upper limb

1. Quadrangular space
Site: In posterior wall of axilla
Borders – superior: subscapularis anteriorly, teres minor posteriorly, inferior: teres major, medial: long head of triceps, lateral: medial shaft of humerus
Contains: Axillary nerve, post circumflex humeral artery and vein

2. Lateral triangular space
Site: In posterior wall of axilla
Borders – superior/medial: teres major, inferior/medial: long head of triceps, lateral: medial shaft of humerus
Contains: Radial nerve and profunda brachii vessels

3. Medial triangular space
Site: In posterior wall of axilla
Borders – superior: subscapularis anteriorly, teres minor posteriorly, inferior: teres major, lateral: long head of triceps
Contains: Circumflex scapular artery

Quadrangular, medial and lateral triangular spaces

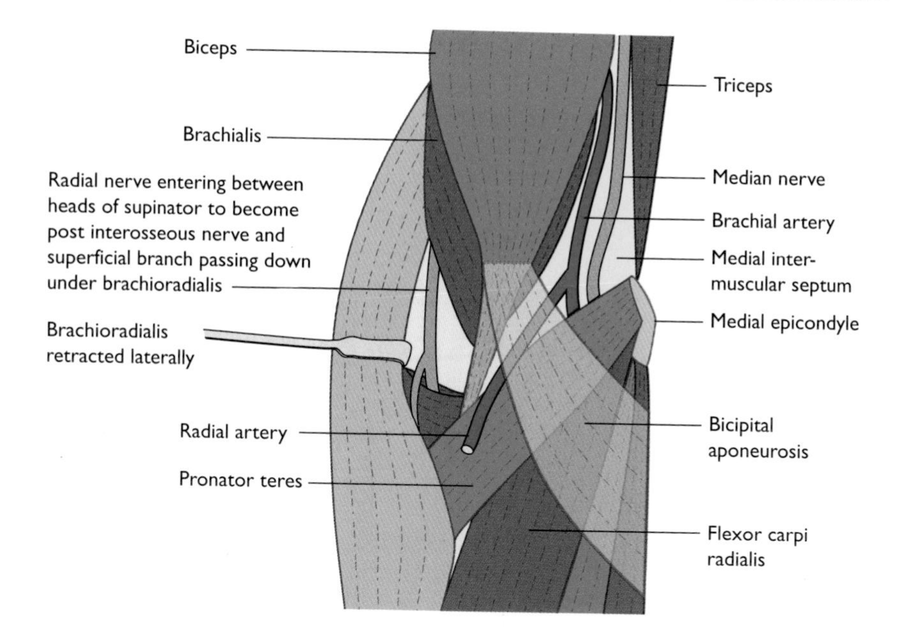

Site: Triangular space in anterior aspect of arm
Borders – superior: intercondylar line, medial: lateral border of pronator teres, lateral: medial border of brachioradialis, floor: brachialis, supinator, roof: fascia
Contains: From lateral to medial: biceps tendon, brachial artery and its accompanying veins, median nerve (i.e. T.A.N. – Tendon – Artery – Nerve), radial and post interosseous and superficial radial nerves under edge of brachioradialis. In fascia of roof are median cubital vein, medial and lateral cutaneous nerves of forearm

Cubital fossa

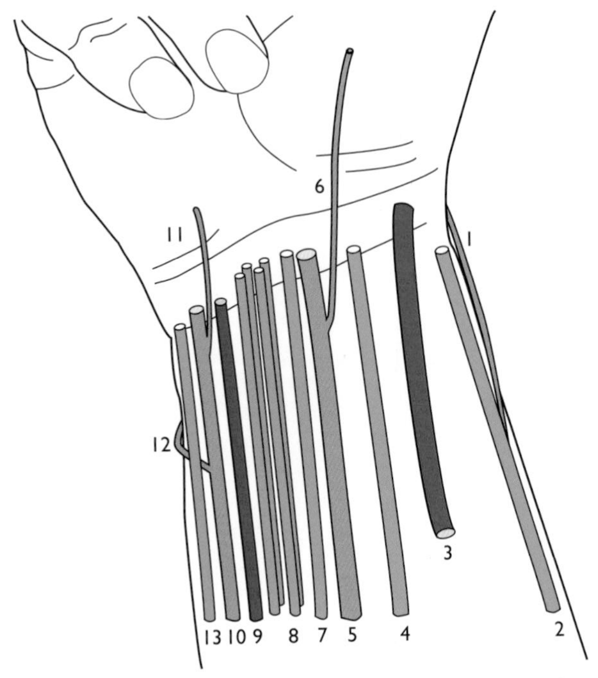

The order of structures across the wrist from the radial side to the ulnar side are:

1 Superficial branch of the radial nerve (emerging posterior to brachioradialis)
2 Brachioradialis
3 Radial artery
4 Flexor carpi radialis
5 Median nerve with its palmar cutaneous branch (6)
7 Palmaris longus
8 Flexor digitorum superficialis (4 tendons)
9 Ulnar artery
10 Ulnar nerve with its palmar cutaneous branch (11) and dorsal cutaneous branch (12)
13 Flexor carpi ulnaris

Ventral aspect of wrist

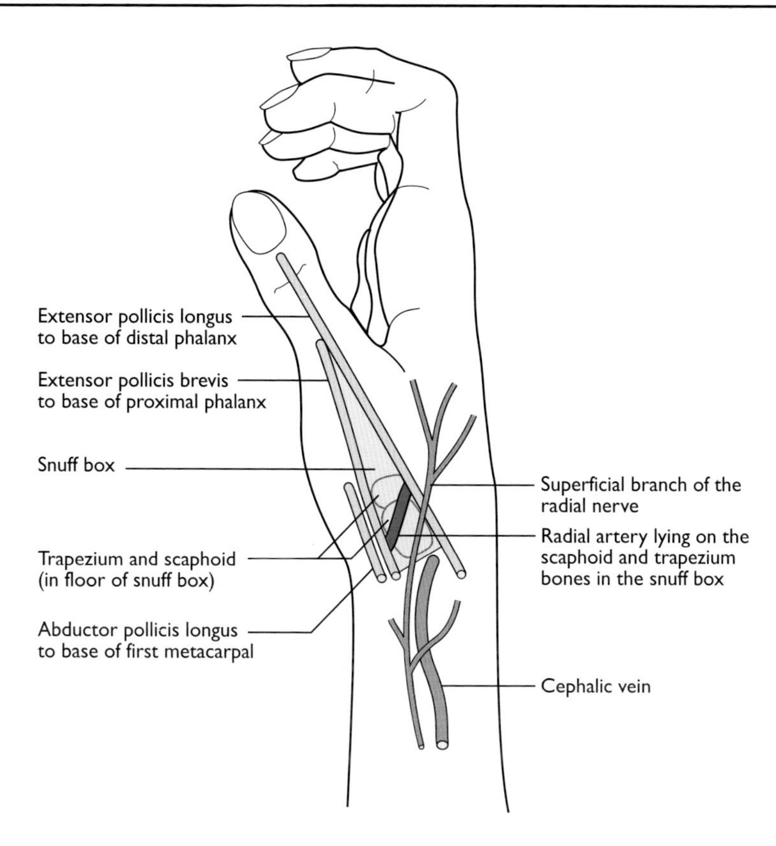

Site: Triangular depression on dorsilateral side of hand formed by tendons of extensor pollicis longus (ulnar side) and extensor pollicis brevis and abductor pollicis longus (radial side)
Contains: Scaphoid, trapezium, radial artery; terminal branches of radial nerve pass over it

Anatomical snuff box

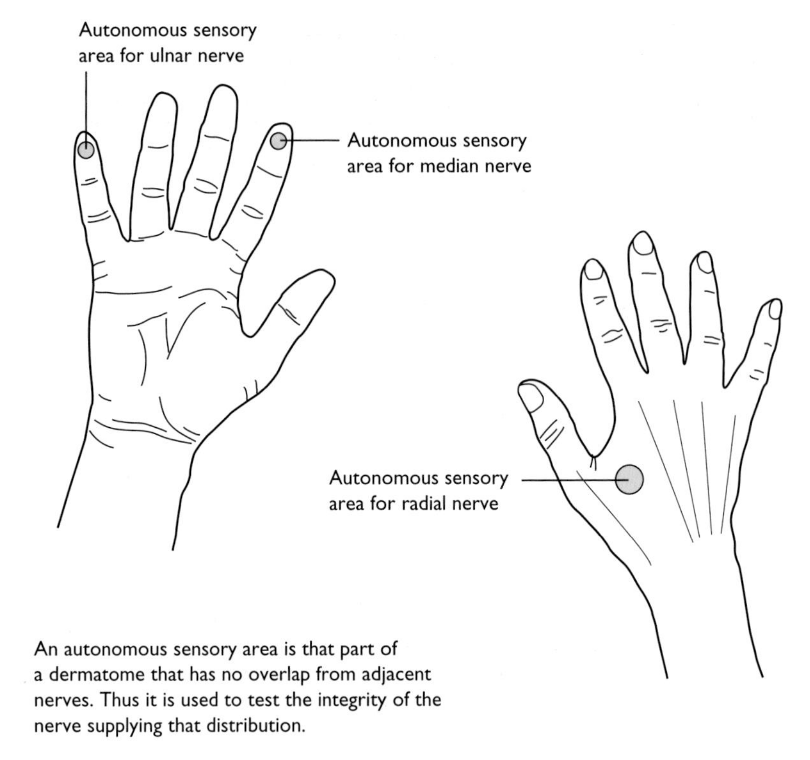

An autonomous sensory area is that part of a dermatome that has no overlap from adjacent nerves. Thus it is used to test the integrity of the nerve supplying that distribution.

Autonomous sensory areas in right hand

Flexor retinaculum
Arises: Tubercle of scaphoid and ridge of trapezium
Inserts: Pisiform and hook of hamate
Inserting into it: Palmaris longus and brevis
Arising from it: Thenar and hypothenar muscles
Superficial to it: Ulnar nerve and artery, palmar cutaneous branches of median and ulnar nerves
Deep to it: See below

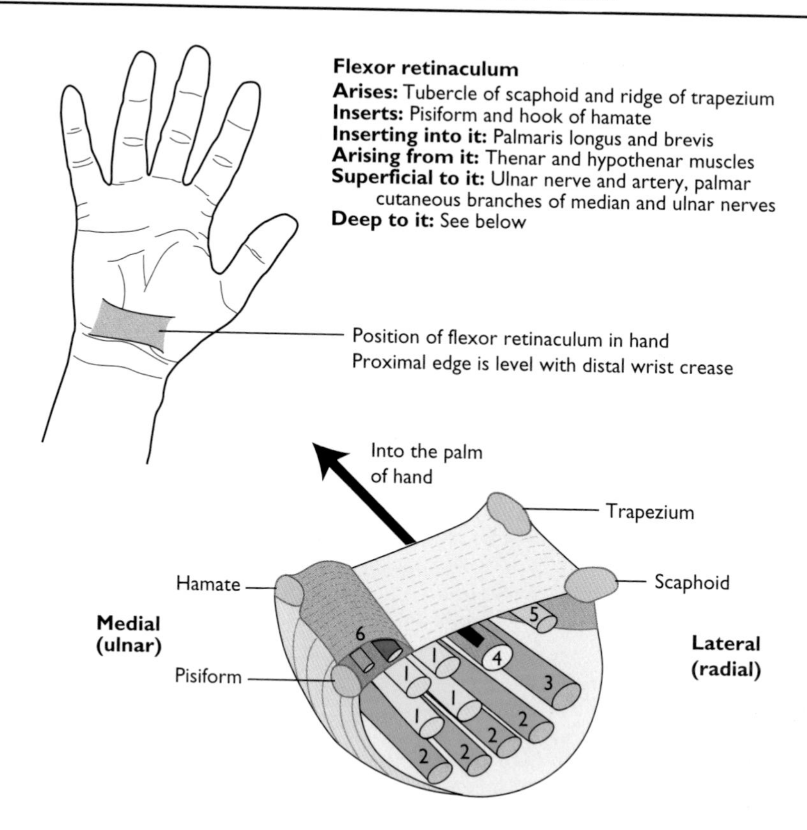

Beneath (deep to) the flexor retinaculum
1 4 tendons of flexor digitorum superficialis
2 4 tendons of flexor digitorum profundus
3 Tendon of flexor pollicis longus
4 Medial nerve
5 Flexor carpi radialis (in its own compartment)

Superficial to the flexor retinaculum but in their own tunnel (Canal of Guyon)
6 Ulnar nerve and ulnar artery

Flexor retinaculum and carpal tunnel

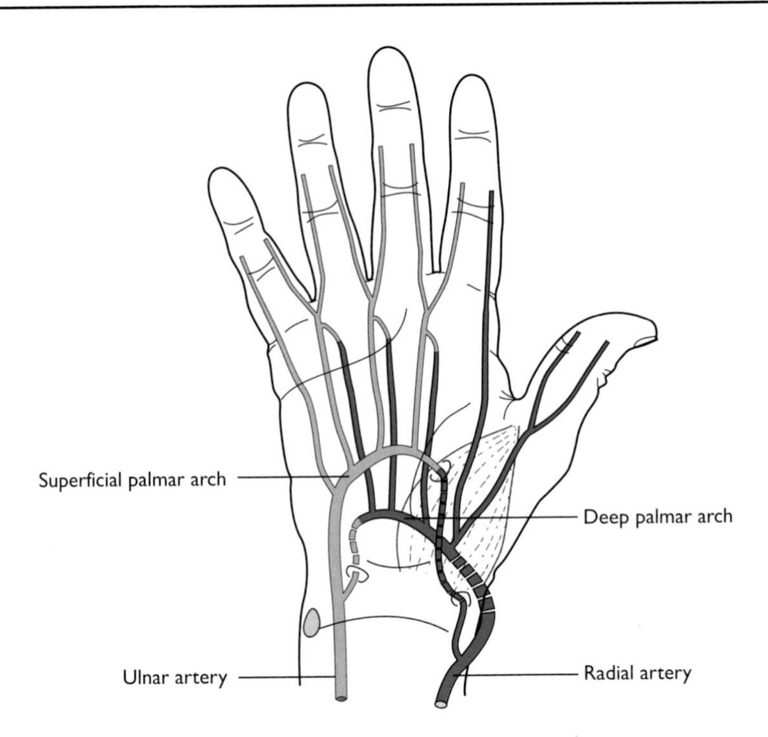

Superficial palmar arch

Deep to the palmar aponeurosis
On a horizontal line at the level of base of outstretched thumb
Branches
- Palmar digital arteries
- Communicating branches with metacarpal arteries of deep arch

Deep palmar arch

From radial artery
Connects to ulnar artery
On a horizontal line one finger's breadth proximal to the superficial palmar arch
Branches
- Princeps pollicis
- Radialis indicis
- 3 palmar metacarpal
- 3 perforating

Palmar arterial arches

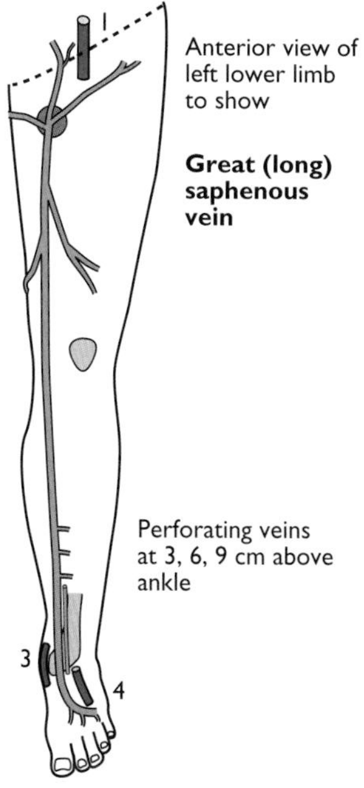

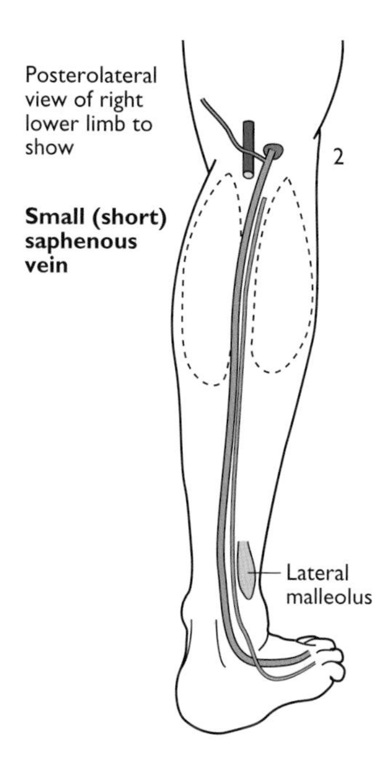

The great saphenous vein commences at the medial side of the dorsal venous arch and passes just anterior to the medial malleolus with the saphenous nerve lying alongside it. It passes a hand's breadth posteromedial to the patella

The small saphenous vein commences at the lateral side of the dorsal venous arch and passes just behind the lateral malleolus. It passes upwards on the posterior aspect of the calf with the sural nerve lying alongside it. It perforates the popliteal fascia and joins the popliteal vein at a variable site

Pulses

1 Femoral pulse in groin at the mid-inguinal point
2 Popliteal pulse deep in the popliteal fossa
3 Posterior tibial pulse behind the medial malleolus, half way between malleolus and heel
4 Dorsalis pedis pulse on the dorsum of the foot, between extensor hallucis longus and extensor digitorum

Veins and pulses in the lower limb

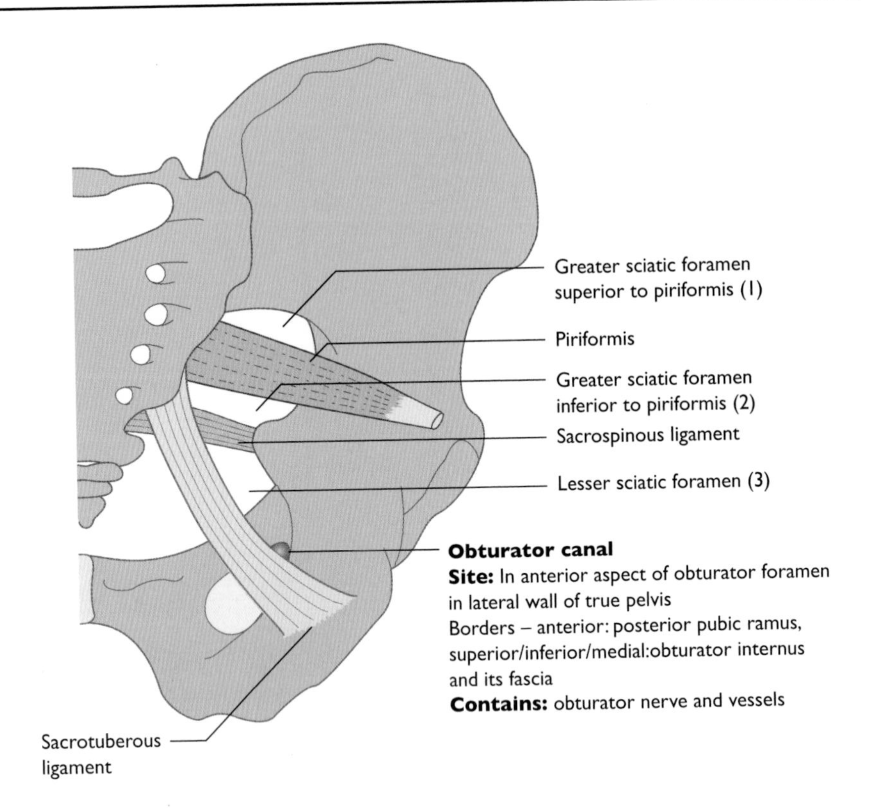

Greater sciatic foramen
Site: In pelvis between greater sciatic notch of ischium/ilium and both sacrotuberous and sacrospinous ligaments. Divided into two sections superior and inferior to piriformis
Passing through superior to piriformis (1): Superior gluteal nerve and vessels
Passing through inferior to piriformis (2): Inferior gluteal nerve and vessels; internal pudendal artery and vein; pudendal, sciatic, posterior femoral cutaneous, perforating cutaneous nerves and nerves to obturator internus and quadratus femoris

Lesser sciatic foramen
Site: In pelvis between lesser sciatic notch of ischium and between sacrospinous and sacrotuberous ligaments
Passing out of it (3): Tendon of obturator internus and internal pudendal vein
Passing into it (3): Nerve to obturator internus, pudendal nerve and internal pudendal artery

Right greater and lesser sciatic foramina and obturator canal
(viewed from behind)

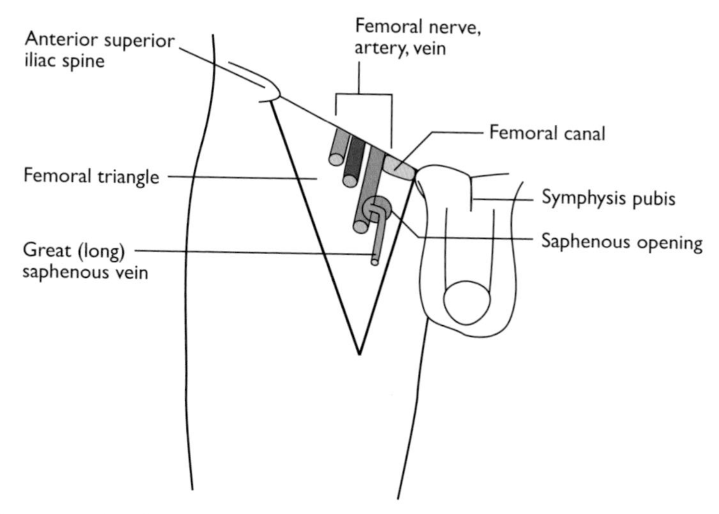

Femoral triangle
Site: In anterior thigh
Borders – medial: medial border of adductor longus; lateral: medial border of sartorius; superior: inguinal ligament; floor: adductor longus, pectineus, iliacus and psoas; roof: fascia lata
Contains: femoral nerve, artery, vein and their branches; deep inguinal lymph nodes

Saphenous opening
Site: In roof of femoral triangle, 3 cm inferior and lateral to the pubic tubercle.
It is a defect in the fascia lata, covered by the cribriform fascia and transmitting the terminal part of the long saphenous vein and lymphatics from the superficial inguinal nodes

Right femoral triangle and saphenous opening

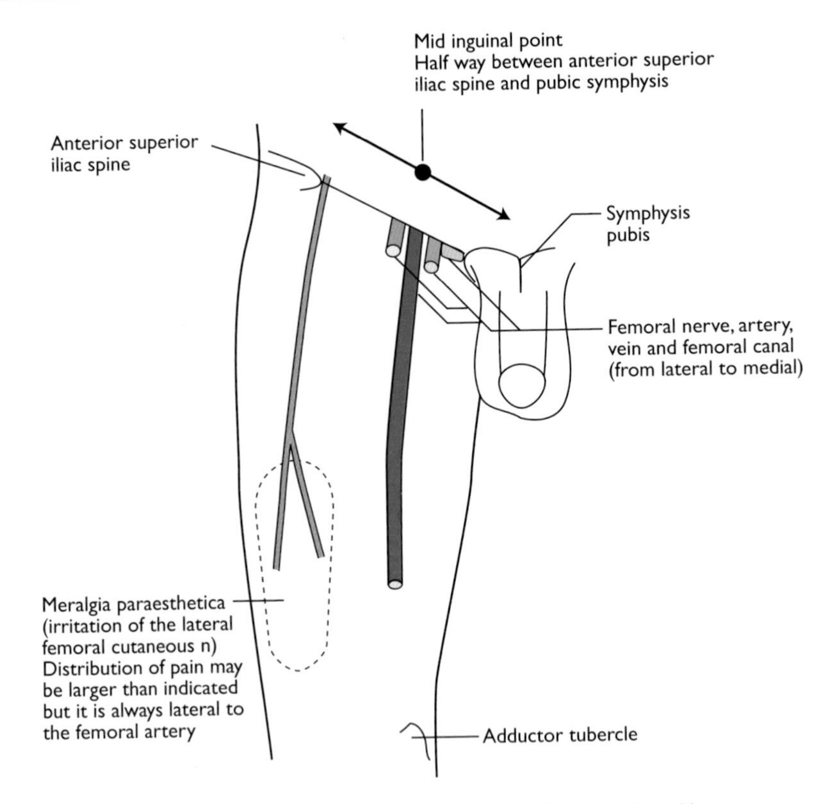

The femoral artery lies 2/3 along a line from the mid inguinal point to the adductor tubercle. It then goes deep to enter the popliteal fossa via the adductor hiatus

Femoral artery in thigh and meralgia paraesthetica

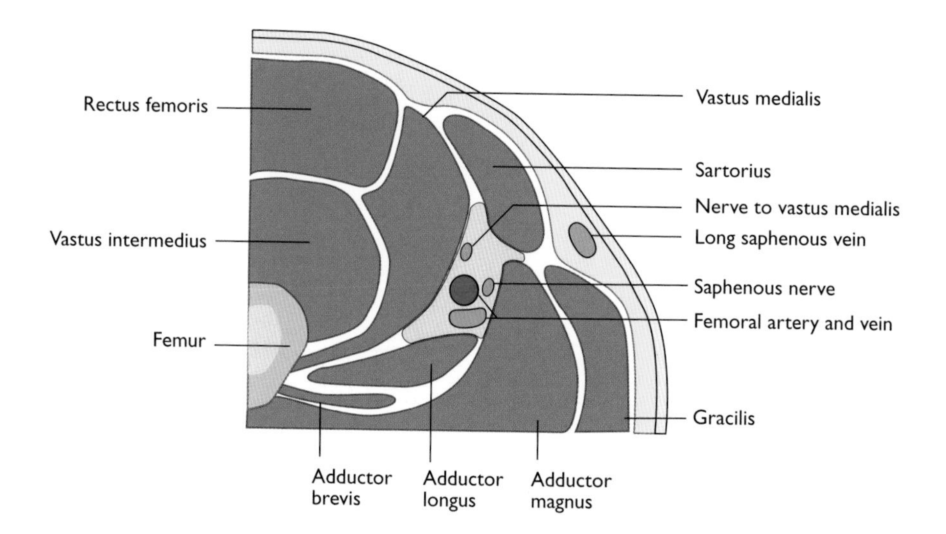

Site: Groove in anterior thigh extending from apex of femoral triangle to hiatus in adductor magnus
Borders – lateral: vastus medialis, medial: adductor longus superiorly and adductor magnus inferiorly, roof: sartorius with its underlying fascia and subsartorial plexus of nerves (These supply skin of medial thigh and have contributions from anterior branch of obturator nerve and medial cutaneous nerve of thigh and saphenous nerve, both from femoral nerve)
Contains: Femoral artery and vein; saphenous nerve; nerve to vastus medialis

Right adductor (subsartorial/Hunter's) canal
(axial section, viewed from below)

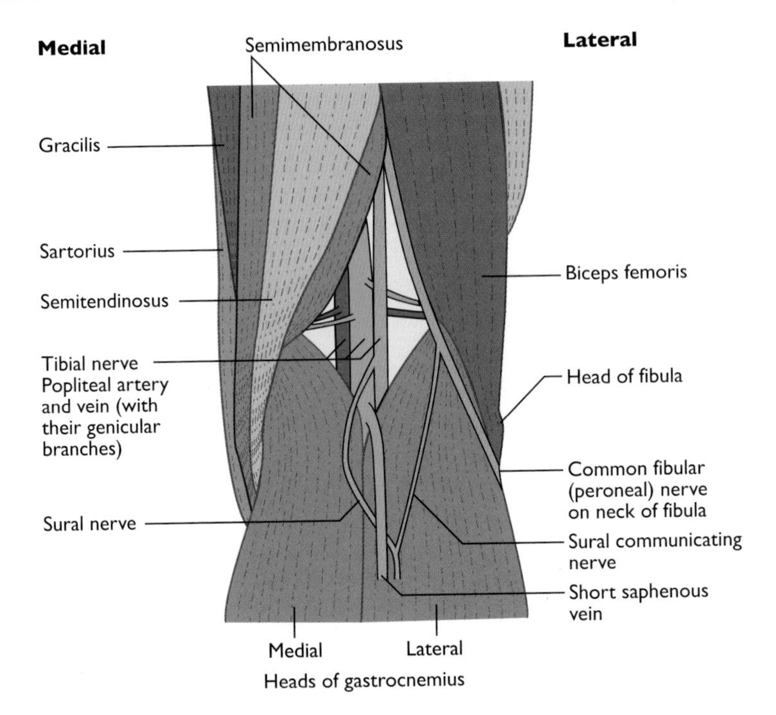

Popliteal fossa
Site: Diamond shaped area posterior to knee
Borders – superior/lateral: biceps femoris, superior/medial: semitendinosus and semimembranosus, inferior/medial and inferior/lateral: heads of gastrocnemius, floor: posterior distal femur, posterior capsule of knee and popliteus, roof: popliteal (deep) fascia
Contains: Plantaris (not shown), popliteal artery and vein and branches; tibial, common fibular (peroneal), sural, sural communicating nerves; lymph nodes and fat. Short saphenous vein and post femoral cutaneous nerve in fascia of roof

Note that the popliteal artery is the deepest structure in the fossa and hence the popliteal pulse is often difficult to palpate

Right popliteal region

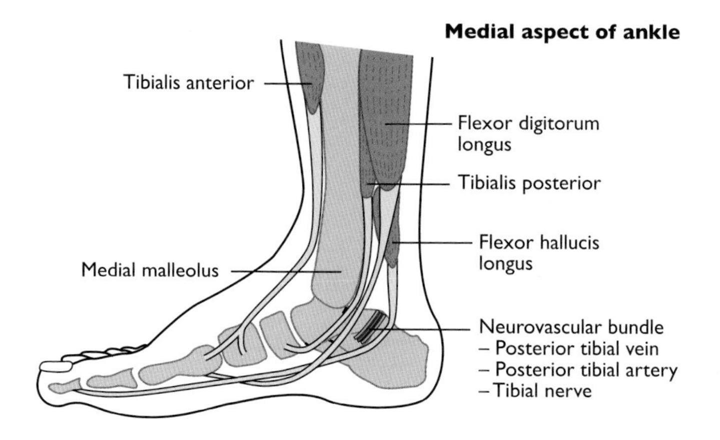

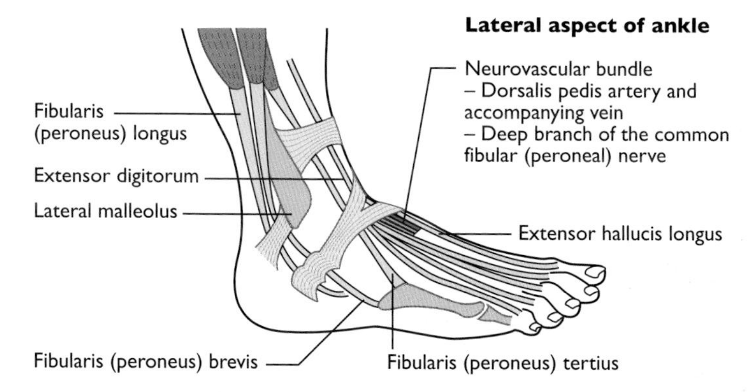

Tendon and neurovascular relationships on medial and lateral aspects of the ankle

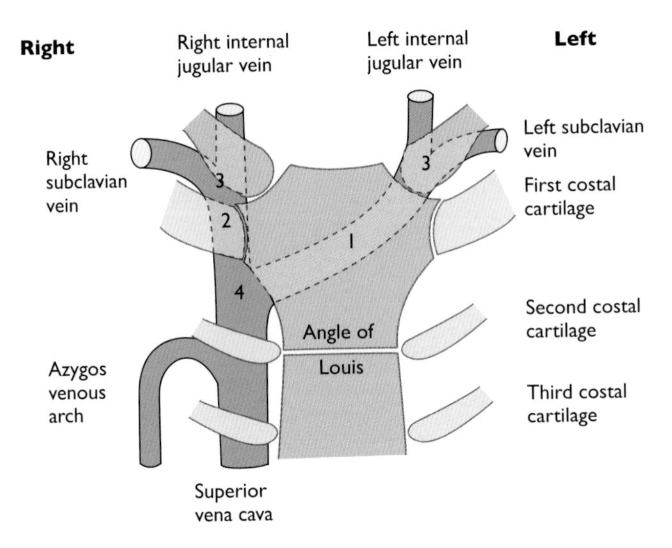

1 Left brachiocephalic vein
2 Right brachiocephalic vein
3 Formation of brachiocephalic veins posterior to the sternoclavicular joints
4 Formation of the superior vena cava in the first right intercostal space, just inferior to the first costal cartilage

Veins in upper mediastinum and lower neck

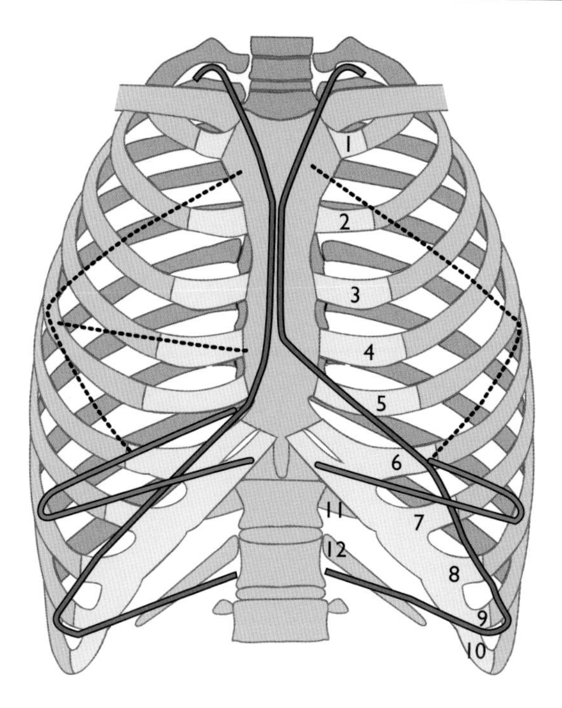

Pleura (red lines)
Start 1 inch (2.5 cm) above mid-point of medial one third of clavicle
Meet in the midline at rib 2
The left side diverges at rib 4 to make room for the heart
The right continues parasternally to rib 6
Both cross rib 8 in mid-clavicular line
Both cross rib 10 in mid-axillary line
Both reach the posterior chest just below rib 12

Lungs (green lines)
Below rib 6, the lungs extend 2 rib spaces less than the pleura in expiration

Lung fissures (black dotted lines)
Oblique – extends from the spine of T3 vertebra around the chest to rib 6 anteriorly. This corresponds approximately to the medial border of the abducted scapula
Horizontal – extends on right from rib/costal cartilage 4 anteriorly to rib 5 in the mid-axillary line

Pleura reflections and lung markings

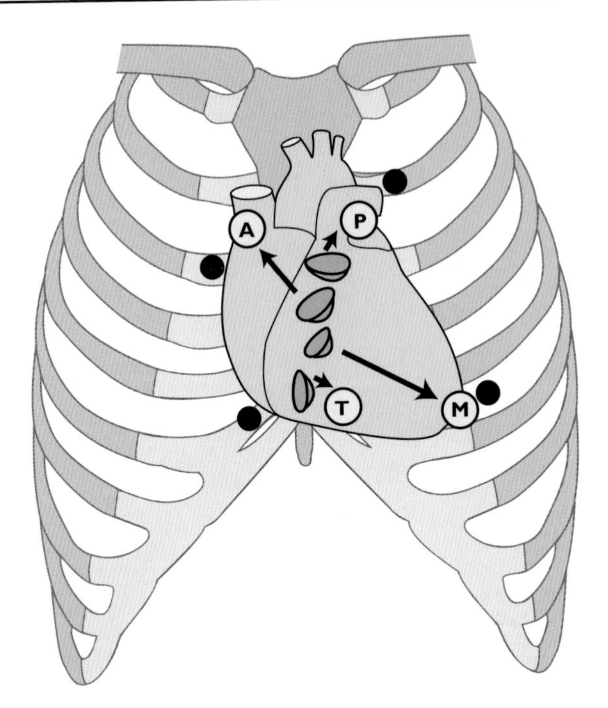

Surface anatomy of a normal sized heart in thorax
Its borders are, anticlockwise starting from upper left:
Left 2nd costal cartilage to right 3rd costal cartilage to right 6th costal cartilage to left 5th intercostal space in the mid-clavicular line (summary 2-3-6-5½)

Auscultation
As the valves open and close they produce sounds that are transmitted in the direction of the blood flow. The 4 valves lie as shown behind the sternum. Sounds are transmitted as follows:
Pulmonary – 2nd left intercostal space, parasternally
Aortic – 2nd right intercostal space, parasternally
Mitral – 5th left intercostal space, midclavicular line (apex)
Tricuspid – over lower sternum

Heart, heart valves and sites of auscultation

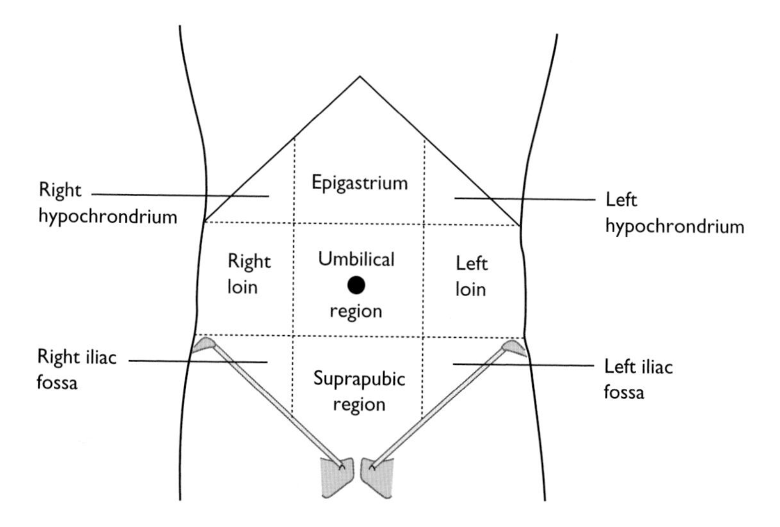

For descriptive purposes, both anatomically and clinically, the abdomen is divided into 9 regions by 2 horizontal and 2 vertical lines. Books tend to define these lines but, in reality and because of human variation in shape and size, the lines are arbitrary.
An alternative method, often used clinically, is simply to divide the abdomen into 4 quadrants, upper, lower, left and right, with 2 loins posteriorly

Regions of the abdomen

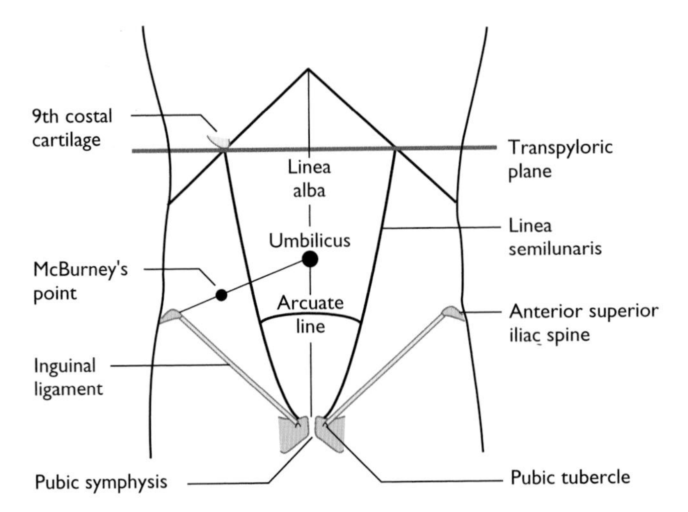

The **linea alba** is in the midline where the 2 sides of the rectus sheath fuse
The **linea semilunaris** is the lateral edge of the rectus sheath on each side
The **arcuate line** is 5–6cm below the umbilicus and is the level where the posterior wall of the rectus sheath finishes and where all 3 anterior wall muscles (external and internal oblique and transversus abdominis) lie anterior to the rectus abdominis
The **transpyloric plane** is a horizontal line which lies half way between the suprasternal notch and the pubic symphysis
McBurney's point indicates the site of the appendix and lies a third along a line between the anterior superior iliac spine and the umbilicus
The **umbilicus** lies at the level of the L3 vertebra

Surface markings on the anterior abdominal wall

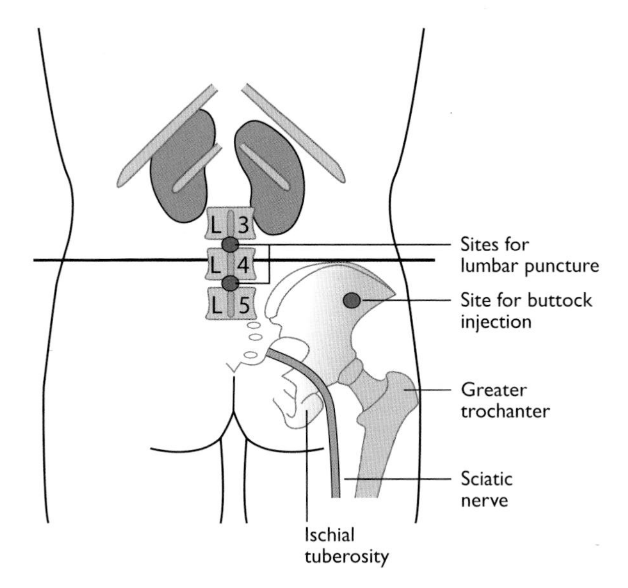

Buttock injection
Intramuscular injections are given in the upper, outer (lateral) quadrant of the buttock to avoid the sciatic nerve which lies in the lower inner (medial) quadrant and passes inferiorly half way between the ischial tuberosity and the greater trochanter of the femur

Supracristal plane
A transverse line along the upper aspects of the iliac crests passing through the spinous process of the L4 vertebra. A useful aid in lumbar puncture for either the L3–4 or L4–5 spaces

Posterior abdomen and back

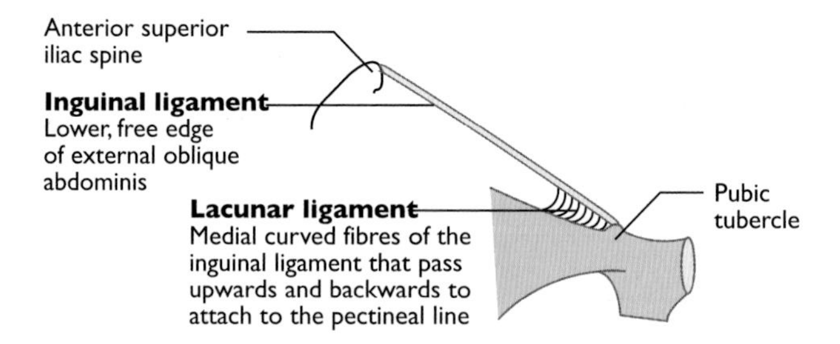

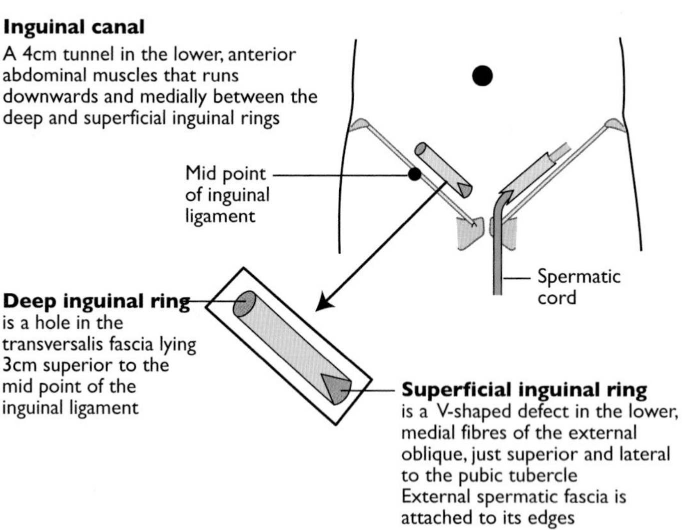

Inguinal ligament
Site: In groin. Details as shown above

Inguinal canal: Details as shown above
Borders – anterior: external oblique abdominis and small portion of internal oblique abdominis laterally, posterior: transversalis fascia and inferior epigastric vessels laterally and conjoint tendon medially, superior (roof): curved fibres of internal oblique and transversus abdominis, inferior (floor): inguinal ligament
Contains: vas deferens/round ligament of uterus (female): testicular, cremasteric and vasal arteries and veins; obliterated processus vaginalis; ilio-inguinal, genital branch of genitofemoral and sympathetic nerves; lymphatics; internal spermatic and cremasteric fasciae

Deep inguinal ring
Site: Details as shown above
Borders – superior and lateral: curved fibres of transversus abdominis, inferior: inguinal ligament, medial: transversalis fascia and inferior epigastric vessels, anterior: curved fibres of internal oblique which will form part of conjoint tendon, posterior: peritoneum. Internal spermatic fascia is attached to its edges. Note that the ilio-inguinal nerve does not pass through it
Contains: All of the above which are in the inguinal canal except the external spermatic fascia and ilio-inguinal nerve

Inguinal ligament and canal

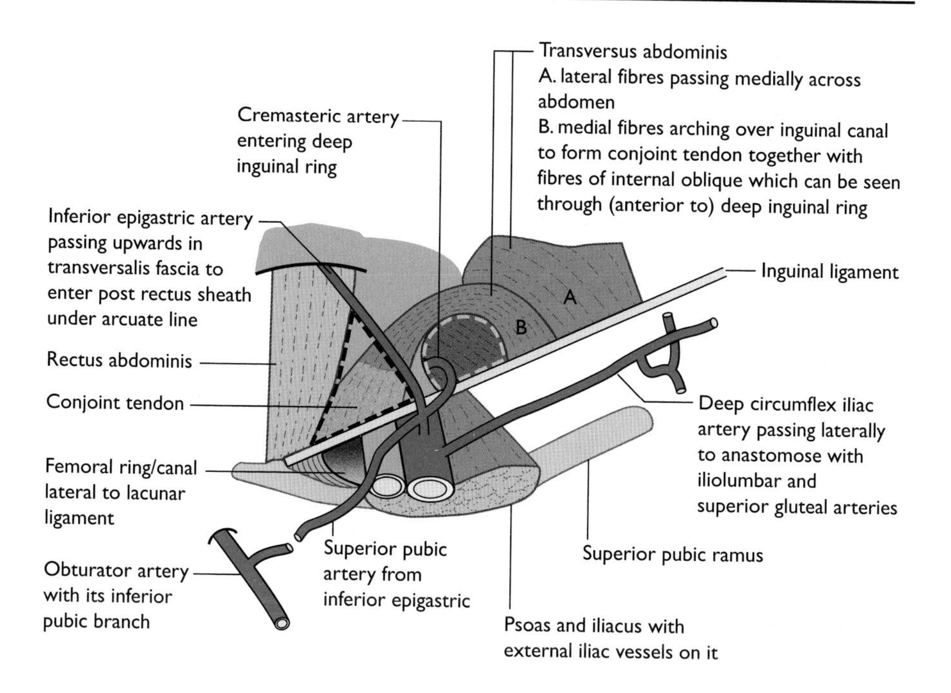

Inguinal triangle (black dashed line)
Site: Posterior aspect of anterior abdominal wall in inguinal region
Borders – lateral: inferior epigastric artery, medial: lateral edge of rectus abdominis, inferior: inguinal ligament, floor: transversalis fascia, conjoint tendon and posterior wall of inguinal canal
Contains: Nil. Site of direct inguinal hernia

Deep inguinal ring (yellow dashed line)(see Inguinal ligament and canal - page 228)

Note: If obturator artery is missing, then superior pubic branch of inferior epigastric takes over. This artery is then called an abnormal (aberrent) obturator artery. Whether or not an abnormal obturator artery is present, the superior pubic branch of inferior epigastric may run anteromedial to the sac of a femoral hernia in the femoral ring. If so, it can easily be damaged during a hernia repair

Right deep inguinal ring and inguinal (Hesselbach's) triangle
(viewed from inside abdomen)

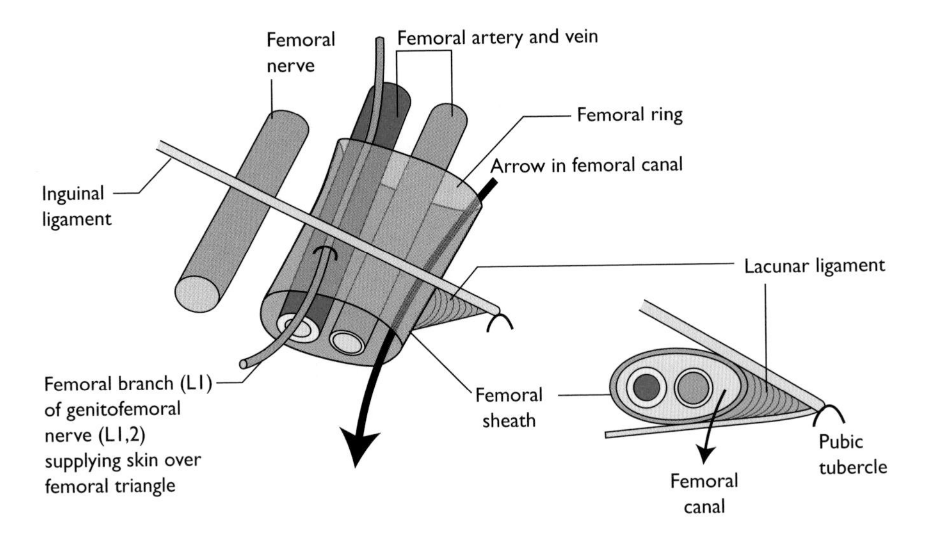

Femoral canal
Site: In lower abdomen, posterior to inguinal ligament
Borders – anterior: inguinal ligament, medial: lacunar ligament, posterior: pectineal ligament and pectineus, lateral: femoral vein
Contains: Lymphatics from lower limb and Cloquet's lymph node either within it or at its superior aspect (femoral ring)

Femoral sheath: Fascia that surrounds femoral canal, femoral artery and vein but not femoral nerve. Its anterior layer is from transversalis fascia and its posterior layer is from fascia over the psoas

Femoral ring: Superior open end of femoral canal

Right femoral canal, sheath and ring

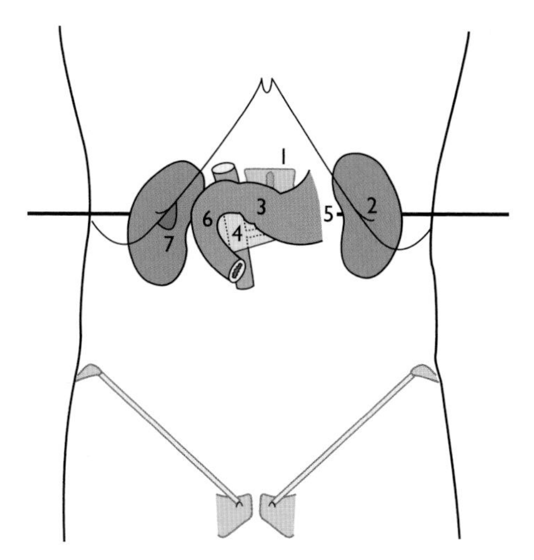

Transpyloric plane
(A horizontal line across the abdomen half way between the suprasternal notch and the pubic symphysis)

The following structures lie **approximately** on this line:

1 L1 vertebral body
2 The tips of the ninth costal cartilages
3 The pylorus of the stomach
4 The neck of the pancreas and origin of the portal vein
5 The hilum of each kidney
6 The second part of the duodenum and sphincter of Oddi
7 The fundus of the gall bladder

Also
The duodenojejunal flexure
The origin of the superior mesenteric artery
The end of the spinal cord

Transpyloric plane

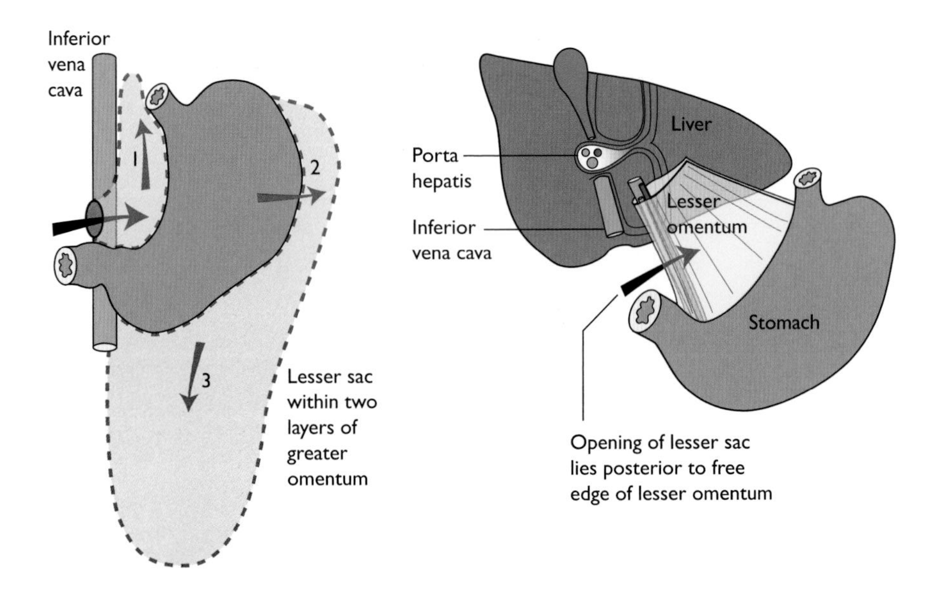

Lesser sac (omental bursa)

Site: An unpaired diverticulum from general peritoneal cavity in upper abdomen opening via the epiploic foramen of Winslow

With rotation of stomach during fetal life peritoneum on its right side is dragged posterior to the stomach and a cavity that it forms (lesser sac) expands as indicated by arrows:

1. Up under left lobe of liver
2. Across to spleen on left side
3. Inferiorly within the 2 layers of greater omentum

Contains: Nil but its peritoneal lining lies against – anterior (from above downwards); inferior/posterior surface of liver, lesser omentum, body and fundus of stomach, greater omentum, inferior: transverse colon, posterior: inferior vena cava, first 2.5cm of duodenum, aorta, coeliac trunk and branches, upper body of pancreas, left suprarenal gland, upper pole of left kidney, superior: caudate lobe of liver, medial/right: opening of sac with inferior vena cava in its post edge, portal vein, hepatic artery and bile duct in its anterior free edge, lateral/far left wall: spleen, attached to stomach anteriorly by gastrosplenic ligament (containing short gastric and right gastro-epiploic vessels) and posteriorly to left kidney by lienorenal ligament (containing tail of pancreas and splenic vessels). Note: This far left wall is effectively the dorsal mesentery of the stomach, the lesser omentum being the ventral mesentery

Lesser omentum: Derives from the ventral mesentery of stomach and runs between liver and lesser curvature of stomach. It is the roof (anterior wall) of lesser sac. In its free edge are portal vein, bile duct and hepatic artery

Lesser sac and lesser omentum

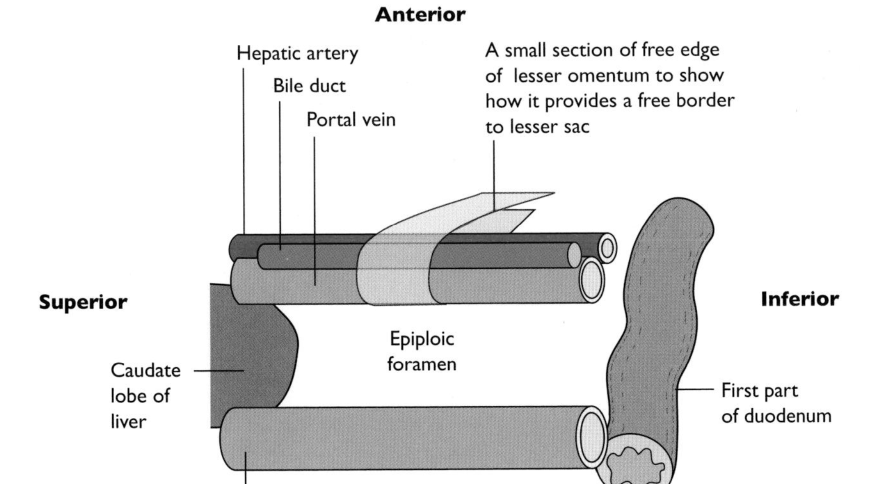

Simplified diagram looking left into lesser sac from right side of upper abdomen

Site: An unpaired opening in upper abdomen approached from right side, leading into less sac of peritoneum
Borders – anterior: portal vein, bile duct and hepatic artery in free edge of lesser omentum, posterior: inferior vena cava, inferior: first part of duodenum, superior: caudate lobe of liver

Lesser sac: (see page 232)

Epiploic foramen

(Foramen of Winslow, aditus to lesser sac)

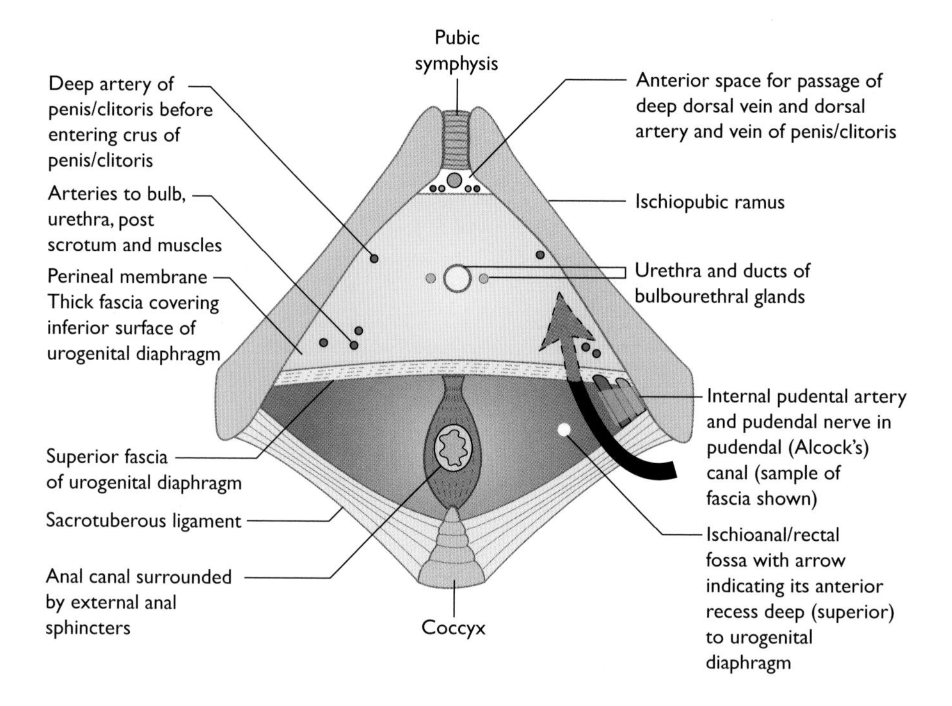

Perineum: That part of trunk that lies inferior to the pelvic floor

Urogenital triangle

Site: In perineum anterior to anal triangle

Borders – lateral: ischiopubic rami, anterior: posterior aspect of symphysis pubis, posterior: transverse line at level of perineal body and ischial tuberosities, floor: urogenital diaphragm

Contains: Deep perineal pouch between superior and inferior fascial layers of urogenital diaphragm (see page 235)

Anal triangle

Site: In perineum posterior to urogenital triangle

Borders – the two sacrotuberous ligaments and the posterior edge of urogenital diaphragm

Contains: Anal canal in midline and an ischioanal fossa on each side

Ischioanal/rectal fossae: Wedge-shaped spaces in anal triangle, lying lateral to anal canal

Borders – medial: anal canal, external anal sphincter, fascia covering inferior surface of levator ani, lateral: ischial tuberosities, fascia over obturator internus, inferior (base/floor): gluteus maximus and buttock skin, apex: junction of levator ani and obturator internus. Has an anterior recess extending forwards on each side deep to urogenital diaphragm

Contains: Fat; pudendal canal (see below), inferior rectal vessels and nerves

Pudendal (Alcock's) canal: A fascial sheath on lateral wall of ischioanal fossa between the opening into perineum of lesser sciatic foramen and posterior aspect of deep perineal pouch

Contains: Internal pudendal vessels and pudendal nerve

Male perineum

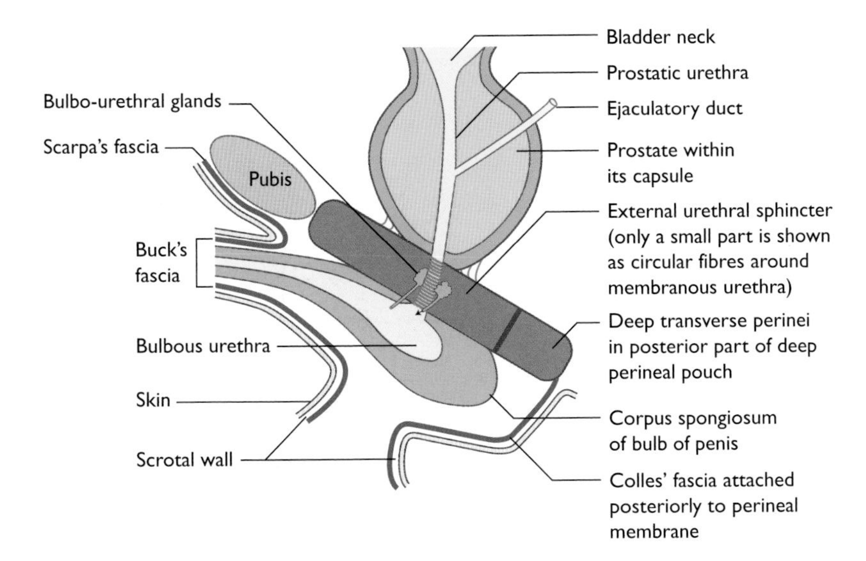

Deep perineal pouch

Site: A space within urogenital diaphragm

Borders – superior: superior fascia of urogenital diaphragm, inferior: perineal membrane, lateral: ischiopubic rami. It has a post free edge and leaves a small gap anteriorly, as shown on page 234, to allow passage of vessels and nerves onto dorsum of penis and clitoris. Perforated by urethra and vagina in female

Contains: Posterior: deep transverse perinei muscles that attach laterally to each ischiopubic ramus, anterior: circular fibres of external urethral sphincter surrounding membranous urethra, lateral: branches of internal pudendal vessels (deep and dorsal to penis and clitoris), branches of pudendal nerve (including dorsal nerve of penis and clitoris). Other contents: two bulbo-urethral glands in males that drain into bulbous urethra below the perineal membrane

Superficial perineal pouch

Site: An area of perineum that lies inferior to perineal membrane

Borders – superior: perineal membrane, inferior: scrotum, penis

Contains in males: All perineal structures below the perineal membrane including the crura of penis, penile and bulbous urethra, perineal body, muscles: superficial transverse perinei, bulbospongiosus (single midline covering bulb), ischiocavernosus (one each side covering crus of penis). All these structures listed above are attached to, and supported by, perineal membrane; branches of vessels and nerves from internal pudendal artery and pudendal nerve

Fascia of perineum

The named fasciae are all deep layers of superficial fascia and are continuous with each other. Colles' is attached to the posterior aspect of perineal membrane and extends in wall of scrotum, Buck's covers the penis, Scarpa's extends down from the abdomen and fuses with the other two

Deep and superficial perineal pouches

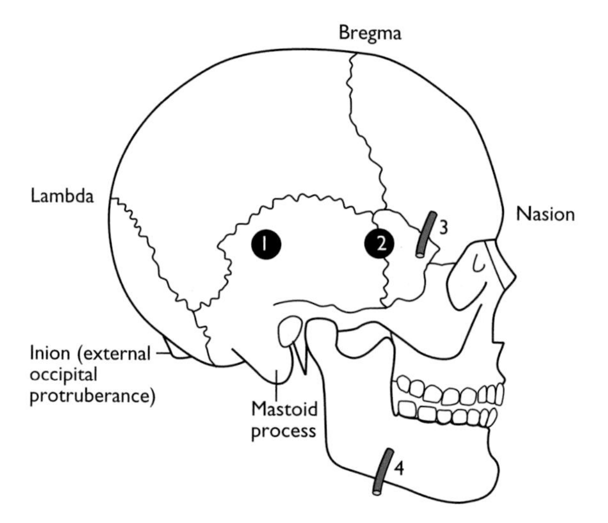

1. Surface marking for the **posterior branch of the middle meningeal artery**: The crossing point of a line vertical from the mastoid process and a line horizontal from the outer canthus of the eye
2. Surface marking for the **anterior branch of the middle meningeal artery**: 3cm above the mid point of the zygomatic arch. (Just inferior to the pterion)
3. Palpable pulse. **Superficial temporal artery**. Over hairless temple
4. Palpable pulse. **Facial artery**. Over mandible, just anterior to the anterior edge of masseter

Skull – lateral view

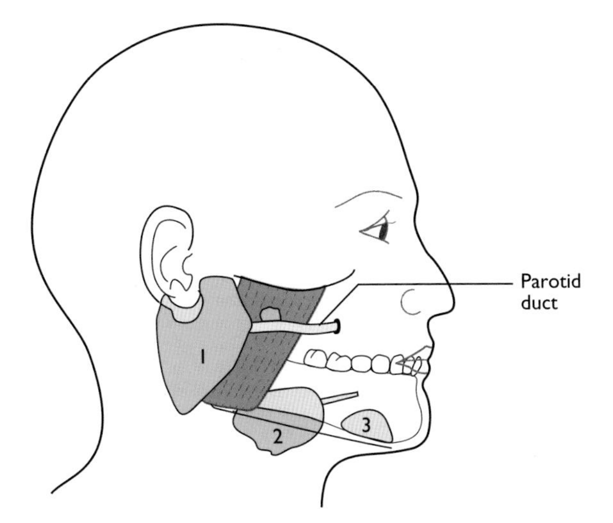

1 **Parotid gland**
The duct is often palpable as it passes lateral to masseter.
It then perforates buccinator and enters the mouth at the level of the second upper molar tooth.
Small accessory parotid glands often enter directly into the duct.
The deep part of the gland extends medially between the ramus of the mandible and the mastoid process

2 **Submandibular gland**
Lies deep to the mandible and extends inferior to it. The duct on each side opens in the floor of the mouth on the summit of its own sublingual papilla

3 **Sublingual gland**
Lies beneath the mucous membrane of the floor of the mouth under the tongue. Its numerous ducts open directly into the floor of the mouth

Salivary glands

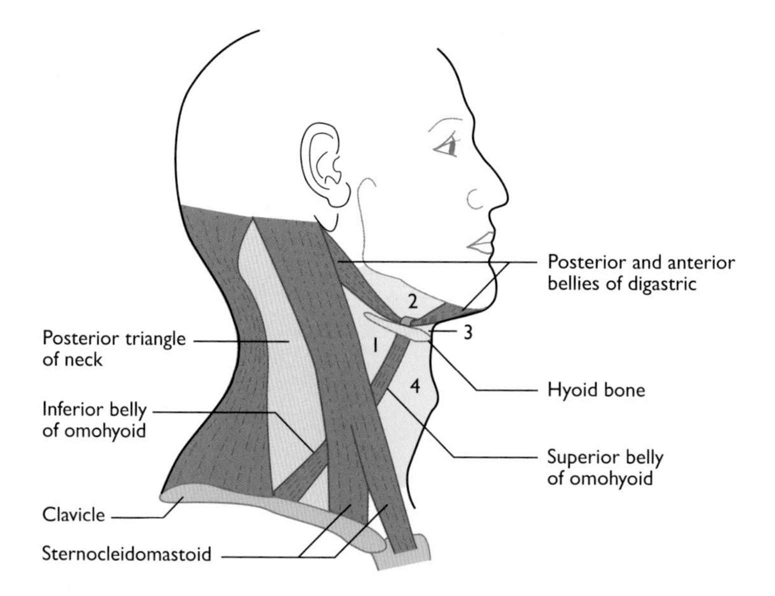

Site: In anterior aspect of neck (shaded blue)
Borders – inferior border of mandible, midline and anterior border of sterncleidomastoid.
Subdivided into carotid (1), digastric (2), submental (3) and muscular (4) triangles
Contains: Muscles – digastric, stylohyoid, mylohyoid, geniohyoid, sternohyoid, omohyoid, thyrohyoid, sternothyroid, platysma. Hyoid bone, larynx, thyroid and parathyroid glands, trachea, oesophagus, submandibular gland and lymph nodes. Arteries – common, internal and external carotids; branches of external carotid; superior thyroid, ascending pharyngeal, lingual, facial; mylohyoid artery (maxillary via inferior alveolar). Veins – internal and anterior jugular, facial. Nerves – hypoglossal, ansa cervicalis, mylohyoid nerve from the mandibular division of the trigeminal nerve via inferior alveolar nerve, vagus and its branches – internal, external branches of superior laryngeal and recurrent laryngeal branches

Digastric triangle: Borders – two bellies of digastric and lower border of mandible.
Floor: mylohyoid and hyoglossus
Carotid triangle: Borders – sternomastoid, posterior belly digastric and superior belly of omohyoid. Floor – middle and inferior constrictors, hyoglossus and thyrohyoid
Submental triangle: Borders – two anterior bellies of digastric and body of hyoid bone.
Floor: parts of both mylohyoid muscles
Muscular triangle: Borders – sternocleidomastoid, superior belly omohyoid and midline of neck

Anterior triangle of the neck

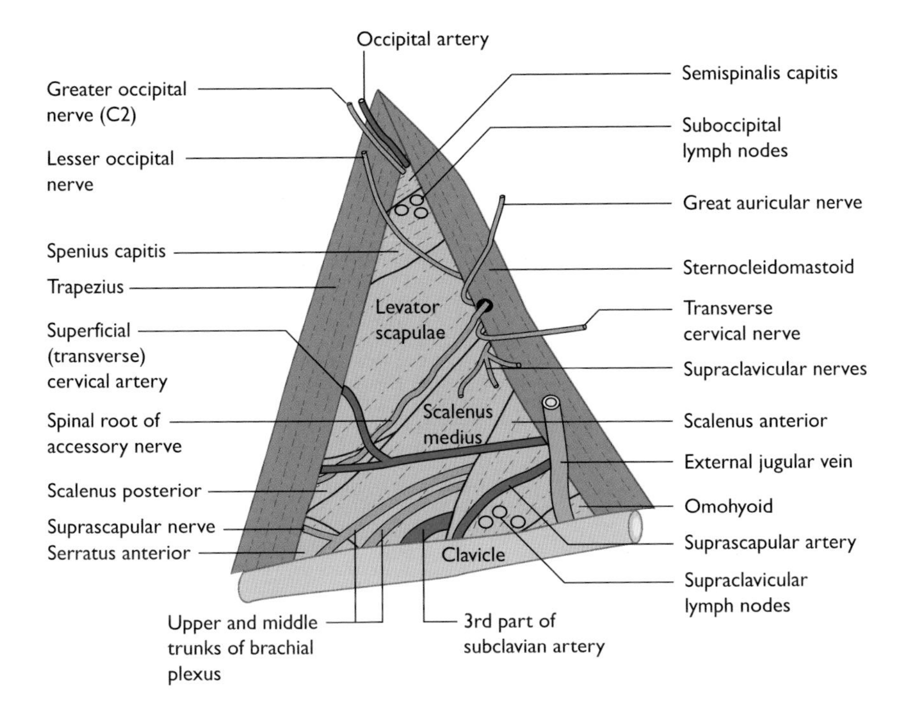

Site: A spiral, triangular shaped space between posterior border of sternocleidomastoid, anterior border of trapezius and middle third of clavicle. Floor: prevertebral fascia over mainly semispinalis capitis, splenius capitis, levator scapulae, scalenus medius and anterior.
Roof: investing layer of deep fascia, platysma, external jugular vein
Contents: Occipital, superficial (transverse) cervical, suprascapular and third part of subclavian arteries; transverse cervical, suprascapular veins; muscular and cutaneous branches of cervical plexus (lesser occipital, great auricular, transverse cervical, supraclavicular nerves); three trunks of brachial plexus; spinal root of accessory nerve; inferior belly of omohyoid; occipital and supraclavicular groups of lymph nodes

Right posterior triangle of neck

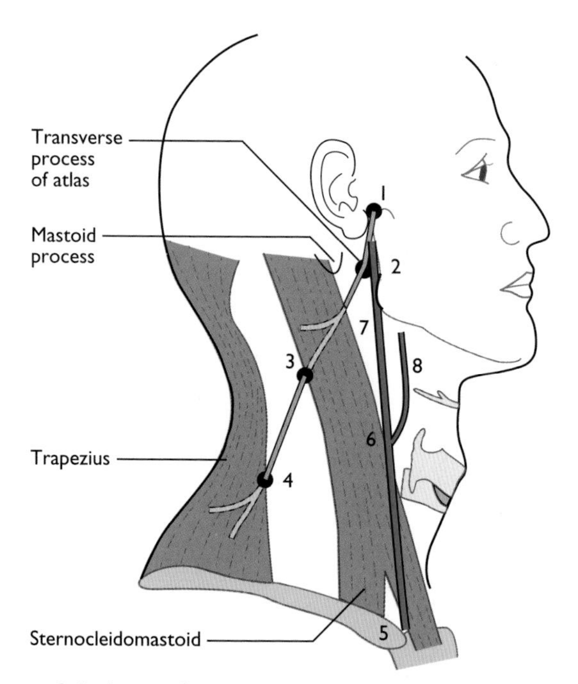

Spinal root of accessory nerve

1 Deep to tragus of the ear
2 Over (lateral to) the transverse process of the atlas
3 Enters the posterior triangle of the neck one third down the posterior border of sternocleidomastoid
4 Enters trapezius one third up its anterior border

Carotid arteries

5 Common carotid artery commences behind the sternoclavicular joint
6 Common carotid artery bifurcates at the level of the upper thyroid cartilage (C4)
7 Internal carotid artery lies behind the neck of the mandible
8 External carotid and its branches

Spinal root of accessory nerve and carotid arteries

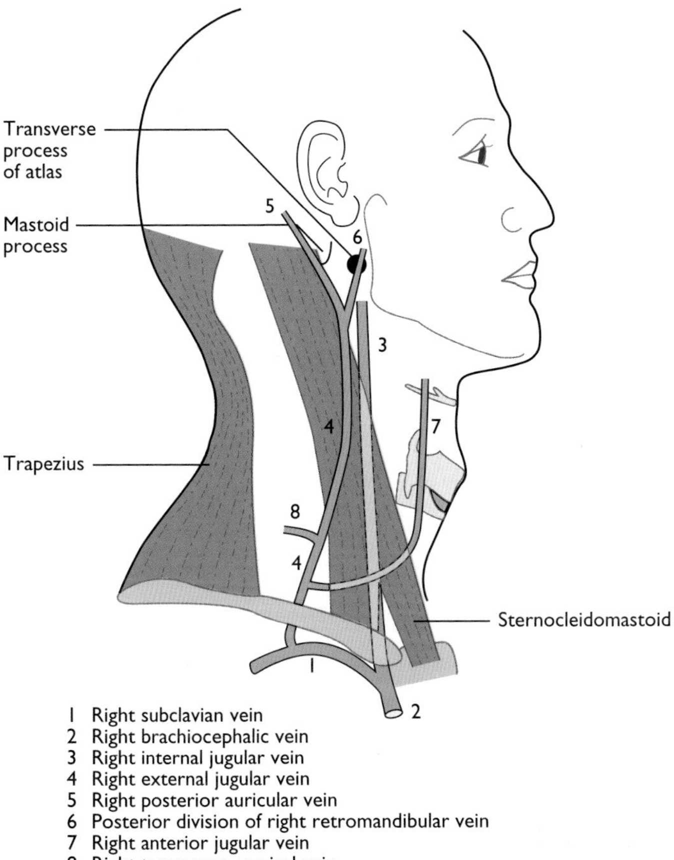

Jugular veins in neck

•

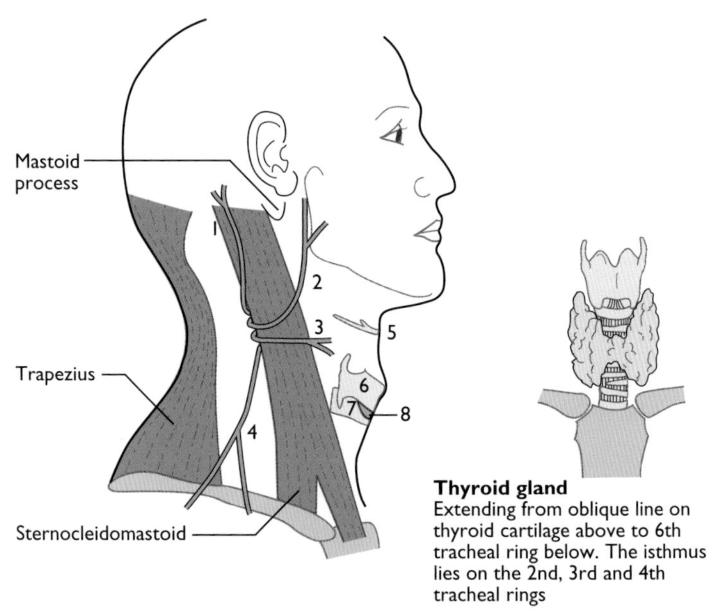

Thyroid gland
Extending from oblique line on thyroid cartilage above to 6th tracheal ring below. The isthmus lies on the 2nd, 3rd and 4th tracheal rings

Cervical plexus

1 Lesser occipital nerve on the posterior border of sternocleidomastoid
2 Great auricular nerve passing to angle of jaw and lower ear. Note that the angle of the jaw is the only part of the face that is not supplied by a branch of the trigeminal nerve
3 Transverse cervical nerve running anteriorly over sternocleidomastoid
4 Supraclavicular nerves passing inferiorly over the clavicle and palpable on it. Pain from the gall bladder can be referred to this dermatome (C4) on the right

Hyoid and larynx

5 Hyoid bone, the greater cornua of which are palpable
6 Thyroid cartilage with its upper anterior prominence (Adam's apple)
7 Cricoid cartilage with the cricothyroid membrane (8) between it and the thyroid cartilage which can be punctured to gain access to the larynx below the cords (cricothyroidotomy)

Cervical plexus, thyroid and larynx